COLON CANCER PREVENTION

Dietary Modulation of Cellular and Molecular Mechanisms

ADVANCES IN EXPERIMENTAL MEDICINE AND BIOLOGY

Recent Volumes in this Series

Volume 462
ADVANCES IN BLADDER RESEARCH
Edited by Laurence S. Baskin and Simon W. Hayward

Volume 463
ENZYMOLOGY AND MOLECULAR BIOLOGY OF CARBONYL METABOLISM 7
Edited by Henry Weiner, Edmund Maser, David W. Crabb, and Ronald Lindahl

Volume 464
CHEMICALS VIA HIGHER PLANT BIOENGINEERING
Edited by Fereidoon Shahidi, Paul Kolodziejczyk, John R. Whitaker, Agustin Lopez Munguia, and Glenn Fuller

Volume 465
CANCER GENE THERAPY: Past Achievements and Future Challenges
Edited by Nagy A. Habib

Volume 466
CURRENT VIEWS OF FATTY ACID OXIDATION AND KETOGENESIS: From Organelles to Point Mutations
Edited by Patti A. Quant and Simon Eaton

Volume 467
TRYPTOPHAN, SEROTONIN, AND MELATONIN: Basic Aspects and Applications
Edited by Gerald Heuther, Walter Kochen, Thomas J. Simat, and Hans Steinhart

Volume 468
THE FUNCTIONAL ROLES OF GLIAL CELLS IN HEALTH AND DISEASE: Dialogue between Glia and Neurons
Edited by Rebecca Matsas and Marco Tsacopoulos

Volume 469
EICOSANOIDS AND OTHER BIOACTIVE LIPIDS IN CANCER, INFLAMMATION, AND RADIATION INJURY, 4
Edited by Kenneth V. Honn, Lawrence J. Marnett, and Santosh Nigam

Volume 470
COLON CANCER PREVENTION: Dietary Modulation of Cellular and Molecular Mechanisms
Edited under the auspices of the American Institute for Cancer Research

Volume 471
OXYGEN TRANSPORT TO TISSUE XXI
Edited by Andras Eke and David T. Delpy

A Continuation Order Plan is available for this series. A continuation order will bring delivery of each new volume immediately upon publication. Volumes are billed only upon actual shipment. For further information please contact the publisher.

COLON CANCER PREVENTION

Dietary Modulation of Cellular and Molecular Mechanisms

Edited under the auspices of the

American Institute for Cancer Research

Washington, D.C.

SPRINGER SCIENCE+BUSINESS MEDIA, LLC

Library of Congress Cataloging-in-Publication Data

Colon cancer prevention: dietary modulation of cellular and molecular mechanisms/
edited under the auspices of the American Institute for Cancer Research.
p. ; cm. — (Advances in experimental medicine and biology; v. 470)
"Proceedings of the Eighth Annual Research Conference of the American Institute for Cancer Research, held September 3–4, 1998, in Washington, D.C."—T.p. verso.
Includes bibliographical references and index.
ISBN 978-1-4613-6861-8 ISBN 978-1-4615-4149-3 (eBook)
DOI 10.1007/978-1-4615-4149-3
1. Colon (Anatomy)—Cancer—Prevention—Congresses. 2. Colon (Anatomy)—Cancer—Nutritional aspects—Congresses. 3. Colon (Anatomy)—Cancer—Molecular aspects—Congresses. I. American Institute for Cancer Research. II. American Institute for Cancer Research. Conference (8th: 1998: Washington, D.C.) III. Series.
[DNLM: 1. Colonic Neoplasms—prevention & control—Congresses. 2. Colonic Neoplasms—pathology—Congresses. 3. Colonic Neoplasms—therapy—Congresses. 4. Diet Therapy—Congresses. 5. Drug Therapy—Congresses. WI 529 C718 1999]
RC280.C6 C637 1999
616.99'434705—dc21

99-048703

ISSN 0065-2598

American Institute for Cancer Research

Proceedings of the Eighth Annual Research Conference of the American Institute for Cancer Research, held September 3–4, 1998, in Washington, D.C.

ISBN 978-1-4613-6861-8

Originally published by Kluwer Academic / Plenum Publishers in 1999
Softcover reprint of the hardcover 1st edition 1999

PREFACE

The Eighth Annual Research Conference of the American Institute for Cancer Research, held in Washington, D.C., September 3–4, 1998, was on the subject "Colon Cancer Prevention: Dietary Modulation of Cellular and Molecular Mechanisms," with participants representing various disciplines interested in this area. One of the speakers provided an appropriate quote from 17th century physician Thomas Adams: "Prevention is better than healing because it saves the labor of being sick," which aptly describes the need for the prevention of cancer.

An overview of normal and abnormal colonic development emphasized that although the typical human colon undergoes 10^{13} cell divisions by age 60, with the associated possibilities for error, relatively few colon tumors develop. Since dietary modulation leads to extremely small changes in colonic cells over a long period, animal models are useful to time, observe, and delineate the events associated with colon cancer.

In the development colon cancer, the inactivation of the adenomatous polyposis coli (Apc) gene is one of the earliest known events. Normally Apc downregulates the cellular protein beta-catenin, but this is lost during cancer development. Beta-catenin may itself be an oncogene; it has a short half-life, but it is stabilized by binding to caherin. Although the Apc is more prevalent in the cell nucleus, the gene shuttles between the nucleus and the cytoplasm.

Short-chain fatty acids, especially butyrate from fermentable dietary fiber, stimulate differentiation, mitochondrial gene expression, and apoptosis. Transcription factors such as the Cdx proteins are also able to control proliferation and differentiation pathways.

Epigenetic mechanisms in colon cancer were reviewed, with special reference to DNA methylation as mediated through dietary folate and methionine. Hypomethylation appears subsequent to the loss of Apc, and several oncogenes are hypomethylated. Folate status is inversely related to colon cancer risk, but high alcohol levels block folate's beneficial effects. Another detrimental factor for colon cancer is chronic inflammation, as associated with release of eicosanoids and cytokines, mediated in part by excess iron and n-6 fatty acids. Various new biological models are being developed to investigate these phenomena. Energy restriction to 60–90% of the *ad libitum* calories in rats reduced tumorigenesis in females given methylnitrosourea. The mechanism appeared to be mediated through the adrenal gland with an increase in corticosteroids which delay cell cycle progression.

Since colon cancer occurs in approximately 50 of every 100,000 people, dietary intervention studies become large, long, and costly, often requiring attention to many factors as vitamins, minerals, fiber, fat, fruit, and vegetable consumption. They still may not provide definitive answers. However, a small intervention trial on the active metabolite from Sulindac, a nonsteroidal anti-inflammatory drug (NSAID) in familial adenomatous polyposis patients showed encouraging results. After two years on the highest dose of the sulfone metabolite that did not cause hepatic toxicity, the intestinal polyps had disappeared. Mechanistic studies indicated cell proliferation decreased and apoptosis was restored.

Epidemiological studies on several kindreds with high cancer risks showed that the risks increased with the number of affected relatives. These familial factors play a role in one third to a half of colorectal cancer cases, with the risks increasing with age. Environmental factors such as dietary habits, physical activity or lack thereof, and alcohol play a large role in sporadic cases of colorectal cancer.

The closing session focused on the balance between cell proliferation and apoptosis in both animal and human studies. Diets which are high in fat (especially corn oil), protein, and fermentable fiber, but low in calcium and vitamin D enhance colonic cell proliferation. Diets with unfermentable fiber and fish oil decreased cell proliferation. Thus, the cellular and molecular studies corroborate the advice of nutritionists—a diet emphasizing fruits and vegetables, fiber, and fish is good for one's health—and certainly one's colon.

CONTENTS

1. Transcriptional Regulation in Intestinal Development: Implications for Colorectal Cancer 1
Peter G. Traber

2. Colonic Cell Proliferation, Differentiation, and Apoptosis 15
Leonard Augenlicht, Anna Velcich, John Mariadason, Michael Bordonaro, and Barbara Heerdt

3. Defects in the Regulation of β–Catenin in Colorectal Cancer 23
Paul Polakis, Matt Hart, and Bonnee Rubinfeld

4. Dietary Lipids, Inflammation, and Colon Cancer 33
Connye N. Kuratko

5. Sulindac Sulfone Induced Regression of Rectal Polyps in Patients with Familial Adenomatous Polyposis 45
G. D. Stoner, G. T. Budd, R. Ganapathi, B. DeYoung, L. A. Kresty, M. Nitert, B. Fryer, J. M. Church, K. Provencher, R. Pamukcu, G. Piazza, E. Hawk, G. Kelloff, P. Elson, and R. U. van Stolk

6. Prevention of Colon Cancer and Modulation of Aberrant Crypt Foci, Cell Proliferation, and Apoptosis by Retinoids and NSAIDs 55
Michael A. Pereira

7. Digested Fiber from Wheat Bran Induces cdk Inhibitors which Block Colon Epithelial Cells in G1 65
Eileen A. Friedman and Jingping Wang

8. Mechanisms by Which Energy Restriction Inhibits Carcinogenesis 77
Henry J. Thompson, Weiqin Jiang, and Zongjian Zhu

9. Dietary Intervention Studies of Colorectal Cancer 85
Arthur Schatzkin

10. The Genetics of Hereditary Non-Polyposis Colorectal Cancer and Non-Polypotic Colon Cancer 95
Päivi Peltomäki

11. Familial Association 99
Randall W. Burt

12. Colonic Cell Proliferation and Apoptosis in Rodent Species: Modulation by Diet 105
Robert S. Chapkin and Joanne R. Lupton

13. Nonsteroidal Anti-Inflammatory Drugs (NSAIDs), Cyclooxygenases, and the Cell Cycle: Their Interactions in Colon Cancer 119
Basil Rigas and Steven J. Shiff

Abstracts 127

Index 163

1

TRANSCRIPTIONAL REGULATION IN INTESTINAL DEVELOPMENT

Implications for Colorectal Cancer

Peter G. Traber

Department of Medicine
University of Pennsylvania
3400 Spruce St.
Philadelphia, Pennsylvania 19104-4283

The molecular mechanisms that direct developmental transitions during organogenesis may be important for understanding the biology of cancers in adult tissues. This manuscript reviews the development of the intestinal and colonic epithelium and demonstrates how an understanding of the mechanisms that regulate transcription of epithelial-specific genes may aid in understanding the complex cellular transitions that occur during development. The caudal-related homeobox genes will be used as an example of how transcription factors that direct developmental processes may be involved in the pathogenesis of colorectal neoplasia.

1. DEVELOPMENT OF THE SMALL INTESTINE AND COLON

The development of the differentiated epithelium of the small intestine and colon is delayed with respect to the gross morphological development of the alimentary tract organs. Throughout most of the processes that result in the anatomic structures (stomach, small intestine, colon), the epithelium remains undifferentiated with similar features along the entire length of the small intestine and colon. Although the processes of epithelial development in the small intestine and colon have many similarities, there are some important differences. Intestinal and colonic development has been studied in detail in a number of experimental animals, in addition to a few descriptive studies in humans. Most mechanistic analysis has been performed in rodents and, therefore, developmental events in humans will be compared with mice. Unlike humans, rodents have short gestational periods and, therefore, the timing of developmental events with respect to birth is different from humans.

Colon Cancer Prevention, edited under the auspices of AICR
Kluwer Academic / Plenum Publishers, New York, 1999.

1.1. Development of the Small Intestinal Epithelium

The mature intestinal epithelium is composed of multiple cell types that arise from a stem cell compartment located in crypts.[1–5] The composition of cells in crypts and on the villus is in a complex equilibrium that undergoes continual renewal while maintaining precise interrelationships. This epithelium develops via a series of transitions. Through week 8 of gestation, the human small intestine is lined by a stratified epithelium composed of 2–6 layers of cells with a relatively undifferentiated phenotype, although the cells have a few poorly developed microvilli.[6,7] Between week 9 and 10 of gestation the epithelium thins to a single layer of columnar cells and villi begin to form in the most proximal portions of the small intestine. This important event in development can be referred to as the endoderm-intestinal transition (Fig. 1). The development of villi proceeds as a wave along the cephalocaudal axis until the entire small intestinal lumen is filled with villi by week 11 of gestation.[6,8] Crypts begin to form in the intervillus regions between weeks 10 and 12 and by week 16 the morphological appearance of the epithelium is very similar to that of the adult.[6–10]

All four of the primary intestinal epithelial cell types appear at approximately the same time that the villus structure of the epithelium is being established. Enterocytes evolve through a series of morphological changes which lead to the adult phenotype during the second trimester.[6,11,12] Goblet cells, both mature and immature, appear between week 8 and 10[13] and are noted in both stratified and columnar epithelium. Paneth cells appear somewhat later between weeks 11 and 12.[13] Enteroendocrine cells appear in the intestinal epithelium at approximately the same time as villi begin to develop between weeks 9 and 10.[14] Therefore, marked changes in the epithelial architecture occur simultaneously with the cell lineage allocation from the undifferentiated epithelial cells.

During this developmental process there is a compartmentalization of proliferating and differentiated cells. At week 8, when the epithelium is stratified, proliferat-

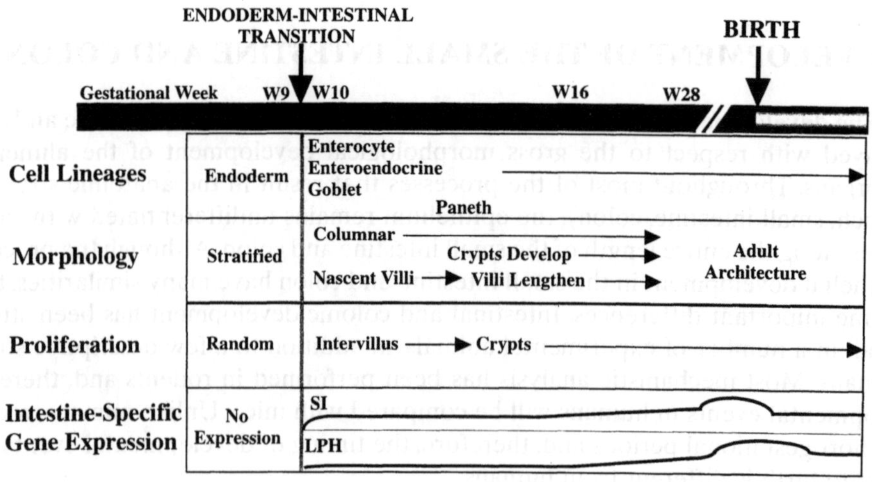

Figure 1. Timeline of events in the development of the human small intestine. See text for description.

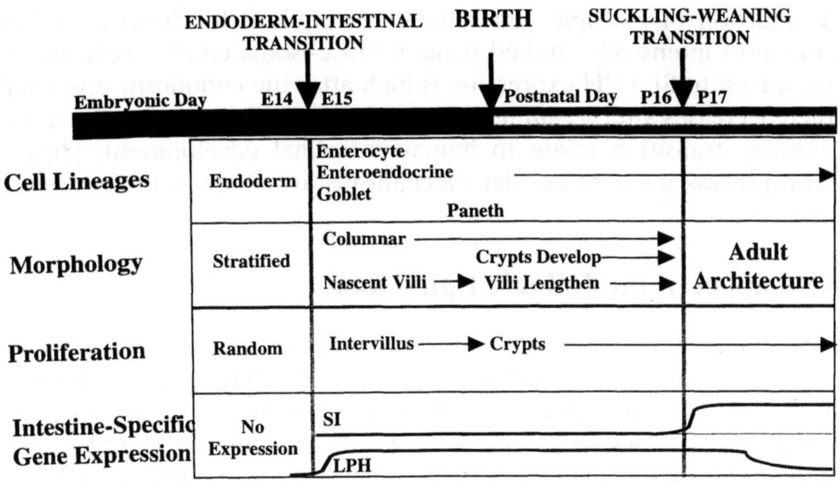

Figure 2. Timeline of events in the development of the mouse small intestine. See text for description.

ing cells are randomly distributed in the epithelium.[15] At the time when nascent villi develop labeled cells are located both on villi and in the intervillus region, but they rapidly localize in intervillus regions. With the development of crypts, the proliferating cells become localized to the crypts. Overall proliferative activity is greatest between weeks 8 and 10 and decreases gradually over the remaining period of gestation.[15] Proliferative activity in mesenchymal tissue, both interstitial cells and the muscularis, parallels the activity in the epithelium.[15–17]

Many intestinal genes are activated during these developmental processes. Two dissacharidase genes, sucrase-isomaltase (SI) and lactase-phlorizin hydrolase (LPH), have been chosen as examples of genes that have complex patterns of expression during development. These have been well examined in both human and rodent intestinal and colonic development. There are many other examples of model genes that have been used to understand intestinal gene regulation during development; the most important of these are the fatty acid binding proteins genes as reported in the elegant studies of Gordon and colleagues.[1,3,18–21] Soon after the endoderm-intestinal transition, both SI and LPH are expressed, reaching a peak just before birth. There is a decrease in expression in the immediate perinatal period, after which the adult levels are established. For LPH, there is a decrease in expression in late adolescence or early adulthood in much of the world's population, but this issue is beyond the scope of this manuscript.

In mice, the endoderm-intestinal transition occurs between gestational day 14 and 15 (E14–E15) (Fig. 2). Therefore, this critical developmental transition occurs after only 25% of the total gestational period in humans, but after nearly 75% of the total gestational period in mice. Consequently, the mucosae of the rodent small intestine and colon at birth are immature, in contrast to the completely developed epithelium of human newborns. Completion of developmental histogenesis and gene expression occurs over the first three weeks of life and is completed by the end of weaning.[22–25] The most dramatic changes occur at the suckling-weaning transition late in the third week of life, between postnatal day 16 and 17 (P16 and P17). At this time there is a marked induction of multiple genes that are required for function of the adult intestine, the most intensively studied being sucrase-isomaltase[22–26] (discussed further below). In contrast to SI, LPH expression is high after the endoderm-intestinal transition and then decreases after weaning. It is not clear how these gene expressions at the suckling-weaning transition relate to human intestinal development. However, the same general processes and molecular mechanisms are likely to be involved in both species.

1.2. Development of the Colonic Epithelium

The development of the human colonic epithelium in many ways recapitulates the development of the small intestine (Fig. 3). At the beginning of the embryonic period (4 weeks) the colon consists of a tube of stratified endoderm surrounded by mesodermal tissue.[27] Between 7 and 8 weeks, three longitudinal ridges arise in the epithelium characterized by thickening of the mucosa but without invagination of mesenchyme. By 9 weeks, mesenchyme has moved to indent the underside of the ridges to form folds in the epithelium which continue to develop into true villiform protrusions into the lumen by week 10. Interestingly, the formation of ridges and folds proceeds from the rectum orad and the ileocecal valve caudad, meeting in the transverse colon. Also at week 10 to 11 there is the first appearance of crypts which appear as protuberances into the mesenchyme between the folds and villi. Secondary lumens are also detected at this time, most likely due to apoptosis, which also contributes to the development of a simple columnar epithelium in a villiform structure. The villi continue to form until the lumen is nearly filled with villi by week 16, at which time the architecture of the fetal colon and small intestine is very similar. In his classical treatise on the subject, Johnson states, "The presence of 'transitory' villi has long been known, being first described by Barth in 1868".[27] Thereafter, the villi gradually regress and they are completely absent by birth. The crypts continue to develop during the fetal period by

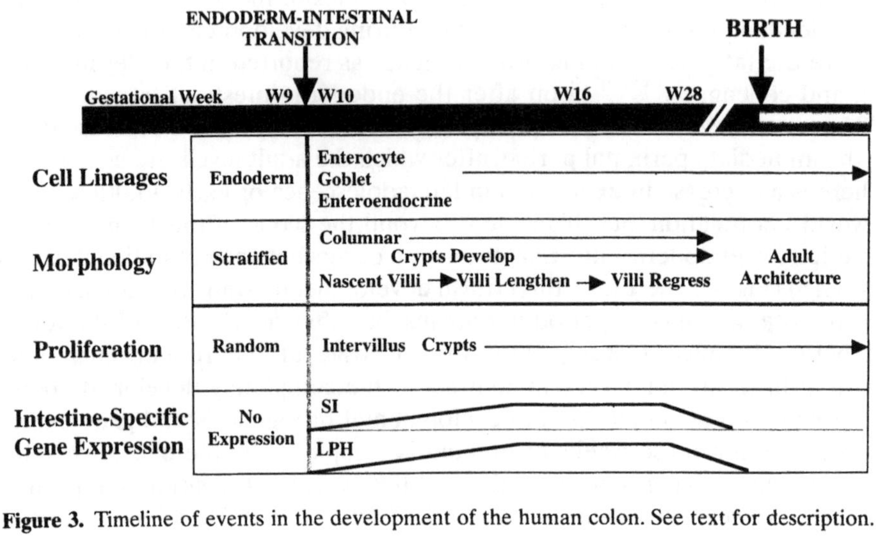

Figure 3. Timeline of events in the development of the human colon. See text for description.

splitting of existing glands from the bottom up. Similar to the histogenesis of the small intestine, this process occurs in a ordered fashion from the proximal to the distal colon. The earliest appearance of goblet cells is between weeks 9 and 10, although appreciable numbers are not seen until week 12, a process that is somewhat delayed from that in small intestine.[28] Unlike the small intestine, the colonic goblet cells rapidly become the predominant epithelial cell lining the fetal colon.[28] Enteroendocrine cells are also first identified between weeks 9 and 10 with increasing numbers up to week 20.[29] Paneth cells are rare in the colon, but accumulate in number in chronic inflammatory conditions.

In addition to the similarity in morphology of the developing colon with the small intestine, there are correlations in the sets of genes that are expressed. Coincident with the transition to a intestinal-type epithelium there is expression of genes that are also expressed in the small intestine including SI, LPH, alkaline phosphatase, aminopeptidase. SI is expressed in the fetal colon beginning at approximately gestational week 11 and it reaches a plateau in week 14 that is maintained until week 28 (Fig. 3).[30] After week 28 there is a gradual loss of SI expression with total absence of SI in the newborn colon.[30,31] The level of expression of SI in the fetal colon is substantial with some studies indicating that it may be as high as the small intestine at similar times of development. Studies using the RT-PCR suggest that there is no expression of SI mRNA in adult human colon,[32] although there is some controversy about this matter.[32–36] Importantly, SI is expressed in neoplastic tissue from the colon, including adenomas and adenocarcinomas.[32–35,37] This may represent a reversion to a fetal type of regulation of the SI gene. The pattern of expression of LPH in the human colon is similar to SI and is not expressed in the neonatal colon. Moreover, as for SI, some colorectal cancer cell lines express LPH and thus simulate the fetal colonic phenotype.

The gross and histologic morphology of the rodent colon during development has many similarities to that of the human colon, with the same timing differences that pertain to small intestinal development (Fig. 4). The development of the rat colon has been better documented than that of the mouse, but the developmental patterns are

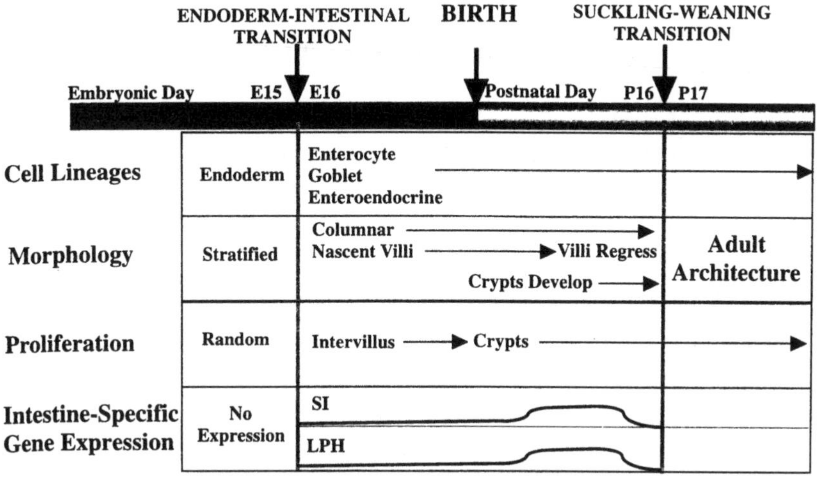

Figure 4. Timeline of events in the development of the mouse colon. See text for description.

similar except for slightly different timing. After the endoderm-intestinal transition in the small intestine, the colonic endoderm undergoes transition with the development of clefts in the stratified epithelium, secondary lumina formed by apoptosis,[38] and then columnar transformation.[39] Goblet cells are found at this early stage of development. Villi develop after birth in the cecum, proximal, and transverse colon which are covered with specialized columnar cells.[39] The distal colon has longitudinal mucosal folds, but never develops a villus structure.[38,39] Villi are diminished in the proximal colon by 10 days after birth and by the third week of life the villi have completely regressed. Crypts develop throughout the postnatal period, often with a branched morphology, and primary and secondary folds in the mucosa develop. In the adult colon, undifferentiated and proliferating cells are located in the mid crypt in the proximal colon and in the base of the cypt in the distal colon.[39,40]

Intestinal gene expression is also noted in the developing rodent colon. SI protein and enzymatic activity are not measurable in the developing colon.[41] However, SI protein is detectable by immunohistochemistry in neonatal rat colon after exposure to exogenously administered corticosteroids.[41] In contrast to the protein, SI mRNA is detected in the neonatal mouse colon when measured using sensitive methods.[42] Low levels of SI mRNA are expressed in the mouse colon at birth, peak between 8 and 10 days and then regress after postnatal day 16.[42] The level of SI expression in the mouse colon at the peak of expression is greater than comparable levels in the small intestine at the same time during postnatal development.[42] However, the levels of SI mRNA in the colon are manyfold less than the expression in the adult small intestine. LPH is expressed in the neonatal rat colon in the same manner as SI.[31,43] Expression in the perinatal rat colon reaches a peak between 2 and 6 days after birth and then decreases to undetectable levels in the third week of life.[41]

Thus, development of the colonic epithelium bears a striking resemblance to that of the small intestine in the earliest, and most crucial, periods of development. The emergence of different cellular phenotypes, the appearance of villi and the development of crypts have the same general temporal and organizational relationships. These similarities, and the ultimate regression of the intestine-like fetal colonic architecture, suggest common mechanisms with either loss of function in the colon epithelium or active repression.

2. MESODERM–ENDODERM INTERACTIONS

The primitive gut tube consists of a stratified endodermal layer of undifferentiated cells surrounded by mesodermal cells, or mesenchyme. In the mature gut, there is a complex interdigitation of endodermal and mesodermal cellular derivatives that have been segregated into multiple lineages. As in other tissues, the interaction between mesenchymal and epithelial cells is crucial for coordinated development of the gut and maintenance of the cellular phenotypes (Fig. 5).

The importance of the interaction between mesoderm and endoderm in organogenesis has been established for many organs in a variety of organisms[44] as well as in intestine.[45–51] One experimental design employed in the analysis of these interactions is to mix the undifferentiated epithelium from one tissue with the mesenchyme of another and observe the resultant differentiated epithelial phenotype. These studies have shown that the requirements for differentiation of epithelia vary from organ to organ. Some epithelia differentiate into the adult type epithelium upon interaction with multiple types of mesenchyme, whereas others are totally dependent on the mes-

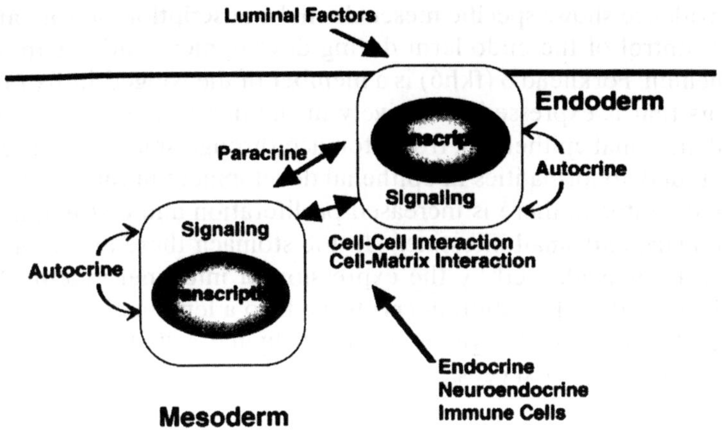

Figure 5. Mechanisms of cellular interactions between mesodermally-derived cells and endoderm that are involved in directing developmental and differentiation events. See text for description.

enchyme of the specific organ of epithelial cell derivation.[44] These organ specific effects are well demonstrated in studies performed in the developing gastrointestinal tract.[47–49] When gastric endoderm is combined with small intestinal mesenchyme, the endoderm develops biochemical characteristics of small intestinal epithelium with expression of intestinal genes.[47] In contrast, when small intestinal endoderm is mixed with gastric mesenchyme, gastric enzymes are not expressed.[47] As an example of the reciprocal relationship, the intestinal endoderm also has a profound effect on the mesenchymal tissue. For instance, skin derived fibroblasts will differentiate into smooth muscle cells when cultured in the presence of intestinal endoderm.[47–49] This establishes the reciprocal nature of the interactions of endoderm and mesenchyme in intestinal development.

Beyond the establishment of the intestinal mucosal architecture, the interaction of epithelial and mesenchymal cells is likely important in maintaining the coordinated cellular relationships in the adult intestinal mucosa. Fibroblasts isolated from postnatal rodent small intestine are able to direct development of cross species heterotopic endodermal tissues (chicken gizzard analage) into small intestine in tissue recombination experiments, whereas fibroblasts from other tissues such as skin allowed development of gizzard epithelium.[47] Therefore, intestinal fibroblasts maintain the ability to produce developmental cues for differentiation of epithelial precursors. These differentiation cues may be important in the maintenance of the intestinal crypt-villus architecture since it has been shown that there is a pericryptal fibroblastic sheath that has proliferative and migratory properties similar to the overlying epithelium.[52]

The mechanisms of the inductive effects between mesoderm and endoderm are dependent on signals exchanged between the two cell types (Fig. 5). Transcription factors in both cell types direct the expressed sets of genes. The set of genes expressed in a given cell type programs the response to signals or initiates signaling cascades that impact on the other cell types. The cells can influence proliferation and differentiation of each other via soluble secreted mediators that can act in autocrine or paracrine fashion, through synthesis of extracellular matrix, or through direct cell-cell interaction. Factors external to the mesenchymal-epithelial unit may affect regulation including endocrine hormones, neuroendocrine mediators, immune cells, or intraluminal factors. However, explant studies have suggested that these external factors may be less important than the intrinsic communication within the mesenchymal-epithelial unit.

Recent evidence shows specific mesenchymal transcription factors are important for regulatory control of the endoderm during development and for maintenance of the adult epithelium. Forkhead 6 (fkh6) is a member of the winged-helix family of transcription factors that is expressed exclusively in the mesenchyme underlying the gut endoderm and intestinal epithelium, from stomach through small intestine.[53] Fkh6 null mice show profound abnormalities in epithelial development in the stomach and small intestine. In both epithelia there is increased proliferation and distortion of the gland structure in stomach and small intestine.[53] In the stomach there are alterations in the phenotype of cells as evidenced by the expression of intestinal mucins. In the small intestine, the increased proliferation in crypts leads to a lengthening of the villi. These data suggest that fkh6 regulates genes in mesenchymal cells that regulate signaling pathways influencing epithelial cells.

3. REGULATION OF INTESTINAL EPITHELIAL GENES AND DEVELOPMENTAL MECHANISMS

During development, and in the adult epithelium, cellular phenotypes are defined by the expression of specific sets of genes in individual cells. The sets of genes expressed in intestinal epithelial cells are principally determined by transcriptional initiation, and the particular set of genes expressed in a single cell type has recently been referred to as the "transcriptome".[54] Intestinal epithelial cell transcriptomes shift in well-orchestrated patterns during developmental, differentiation and adaptive processes of the intestinal mucosa. Thus, the molecular mechanisms that regulate transcription of cellular gene sets form the foundation for understanding the mechanisms of developmental and differentiation events.

We have examined the regulatory mechanisms of the sucrase-isomaltase (SI) gene in an attempt to understand intestinal gene expression during development and differentiation. The SI gene is expressed during mouse small intestinal development in a pattern that mirrors several critical developmental transitions.[42,55] SI mRNA is first detected just after the endoderm-epithelial transition in the newly formed columnar cells, but the level of expression can only be detected using the most sensitive methods of mRNA measurement and SI protein is not detectable.[42] SI expression remains at very low levels until the third postnatal week when the pups are in the process of being weaned from mother's milk to solid food. At this suckling-weaning transition there is a marked induction of SI mRNA expression to levels equivalent to those in adult epithelium.[42,55] Multiple other genes are induced at this transition and there are changes in the morphology of the epithelial cells.[24,25] In the adult epithelium SI mRNA and protein are expressed predominantly in differentiated cells located on the villus. In the migration of epithelial cells from the base of crypts to villus tips, SI is first expressed in detectable levels in cells located in the upper crypt region and is expressed in high levels through the mid-villus with diminished levels in cells at the villus tips. Thus, complex patterns of SI expression are maintained in the continually renewing adult epithelium.

3.1. Understanding the Mechanisms of SI Gene Expression Using Transgenic Mice

Gordon and colleagues pioneered the use of transgenic mice to examine the regulatory regions of intestinal genes and their studies of fatty acid binding protein genes have provided much insight into regulatory mechanisms.[1,3,18–21] We have used transgenic mouse experiments to show that the pattern of SI gene expression during development and in the mature epithelium is regulated largely at the level of gene transcription via multiple functional cis-acting DNA elements.[42,55,56] Although these experiments indicate that the full complexity of the regulation of SI gene transcription remains to be described, important principles have been uncovered that will guide additional studies. Most importantly, the elements necessary to direct intestinal epithelial cell-specific expression appear to be embodied in a 201 nucleotide, evolutionarily conserved 5′-flanking region of the gene.[42] Nucleotides –201 to +54 of the mouse SI gene direct expression of a human growth hormone reporter gene specifically to the intestinal epithelium. The transgene is expressed at high levels in villus associated enterocytes appearing at the time of the suckling-weaning transition, similar to the pattern of expression of the endogenous gene. Thus, to a remarkable degree this short promoter was able to recapitulate the pattern of SI gene expression in enterocytes.

3.2. Transcriptional Elements and Cognate DNA Binding Proteins in the SI Promoter

Transcriptional initiation is an intricate biochemical process that involves the interaction of core nuclear machinery (basal transcriptional apparatus) with cell-specific DNA binding proteins. Co-adaptor proteins serve to link the specific DNA binding proteins to portions of the basal apparatus. The coordinated assembly of a complex of many proteins at the transcriptional initiation site of a gene leads to modification of chromatin structure and initiation of RNA synthesis by RNA polymerase II. DNA regulatory elements and associated DNA binding proteins have been carefully assessed within the evolutionarily conserved promoter region of the SI gene.[57] There are at least three groups of transcriptional proteins involved in SI promoter transcription including hepatocyte nuclear factor I (HNF1),[58] caudal -related homeodomain proteins (Cdx),[59,60] and nuclear proteins that interact with a GATA binding site which are as yet poorly characterized.[57] HNF1α and HNF1β proteins interact with two elements in the human SI gene promoter, SIF2 and SIF3.[58] HNF1α transactivates transcription of the SI promoter whereas HNF1β does not activate transcription, suggesting that relative levels of the two proteins may modulate transcriptional activation.[58]

The caudal-related homeobox genes Cdx1 and Cdx2, both of which are expressed in intestinal epithelial cells, bind to two closely apposed binding sites just upstream of the TATA box of the gene and activate transcription of the promoter.[60] These transcription factors will be discussed below.

4. CAUDAL-RELATED HOMEOBOX (CDX) GENES AND INTESTINAL DEVELOPMENT

Following the discovery that Cdx genes were important for transcriptional regulation of an intestine-specific gene, our laboratory and others showed that it was expressed only in the small intestine and colon of adult mice. These findings raised the question of whether Cdx genes played a more generalized role in intestinal epithelial cell development and differentiation.

Several studies in intestinal cell lines have demonstrated the importance of Cdx proteins on the regulation of differentiation and proliferation. Expression of Cdx2 in the rat intestinal epithelial cell line IEC-6 resulted in profound effects on proliferation and differentiation.[61] Soon after induction of Cdx2 expression there was a delay in cell growth for several days followed by a period of exponential growth to a density that exceeded that of the cells that did not express Cdx2. The confluent IEC6 cells expressing Cdx2 underwent a marked morphological change forming a columnar epithelium with tight junctions, microvilli formation, and lateral membrane interdigitations. Additionally, the differentiated cells expressed LPH and SI mRNAs, molecular markers of the differentiated phenotype. Thus, expression of Cdx2 in IEC6 cells directed a remarkable degree of morphological and molecular differentiation.

The effect of expression of Cdx genes has also been examined in colorectal cancer cell lines. Caco2 cells overexpressing Cdx2 showed increased expression of SI, E-cadherin, APC, β4, HD1, and the γ2 chain of laminin 5.[62,63] Another group of investigators reported the effects of expression of Cdx1, Cdx2, and a combination of both in the colon cancer cell line HT-29.[64] Stable lines expressing Cdx2 were found to proliferate more slowly, had decreased growth in soft agar, and formed smaller tumors in nude mice.[64] In contrast, stable lines expressing Cdx1 did not have an alteration in the pattern of proliferation. Therefore, the studies of manipulating the expression of Cdx proteins in cell lines have revealed that these proteins are indeed important for the regulation of cellular proliferation and differentiation.

Knockout mice have been generated for both Cdx1 and Cdx2. Cdx1 null mice are live born and survive until adulthood.[65] Examination of their skeletal structure shows an anterior homeotic shift demonstrating that expression of Cdx1 in the early embryo is important for patterning of the axial skeleton. A detailed report on the small intestine and colon was not provided in this initial publication, although the intestinal tract was grossly intact and must have been functional since the animals survived. Cdx2 null mice are not viable and die in utero before implantation.[66] However, the Cdx2 heterozygous mice (+/–) had an anterior homeotic shift in the axial skeleton that was nearly identical to that reported for the Cdx1 null mice. Additionally, the small intestine and colon developed normally in the Cdx2(+/–) mice, but several weeks after birth they developed colonic tumors that showed evidence of squamous metaplasia. The expression of Cdx2 proteins was completely lost in the tumor cells, possibly on the basis of epigenetic silencing of the normal allele. These results show that continued expression of Cdx2 is necessary to maintain the normal intestinal phenotype.

5. DEVELOPMENTAL TRANSCRIPTION FACTORS AND INTESTINAL NEOPLASIA

Because of the evidence that Cdx proteins are involved in the regulation of proliferation and differentiation in the intestinal and colonic epithelium, several investigators have examined the expression of Cdx genes in human colorectal tumors. Expression of CDX2 mRNA[67] and protein[68] is decreased in a large percentage of colorectal cancers. This finding is quite interesting and potentially important in the pathogenesis of the cancer in light of the findings in the Cdx2 heterozygous knockout mice. In addition, it has recently been described that colon cancers can have mutations in both alleles of the CDX2 gene.[69] Expression of CDX1 mRNA[67] and protein is also decreased in the majority of colorectal cancers. In our studies of immunohistochemistry of CDX1 proteins, there appears to be a graded decrease in immunoreactivity between normal, adenomatous and cancer cells.[70] Thus, these data show that there are marked decreases in the expression of both CDX1 and CDX2 in colorectal cancers.

These are intriguing findings that may indicate that mutation or loss of expression of CDX genes may play a role in the pathogenesis of colorectal cancer. Much more work must be done to determine the relative importance of these finding for colorectal cancer. However, it does demonstrate the power of developmental mechanisms for generating hypotheses about cancer pathogenesis.

6. SUMMARY

Deciphering the complex mechanisms of intestinal epithelial development will require multiple cell and molecular approaches in both in vitro and whole animal systems. Additionally, the use of model organisms such as *D. melanogaster*, *C. elegans*, and zebrafish will help describe paradigms that may be investigated in mammals as well as serve as test systems for findings from mammals. This manuscript reviewed only one approach to understanding intestinal development. However, the Cdx story and the information to be mined from an understanding of SI gene transcription is not at an end. As the other pieces of the transcriptional puzzle of the SI gene are assembled there will be new information to generate hypotheses on the relationship of transcriptional mechanisms to cancer pathogenesis.

REFERENCES

1. Gordon, J.I. Intestinal epithelial differentiation: New insights from chimeric and transgenic mice. *J. Cell. Biol.* 108:1187–1194, 1989.
2. Gordon, J.I. and Hermiston, M.L. Differentiation and self-renewal in the mouse gastrointestinal epithelium. *Curr. Opin. Cell. Biol.* 6:795–803, 1994.
3. Gordon, J.I., Schmidt, G.H., and Roth, K.A. Studies of intestinal stem cells using normal, chimeric, and transgenic mice. *FASEB J.* 6:3039–3050, 1992.
4. Potten, C.S. and Loeffler, M. Stem cells: Attributes, cycles, spirals, pitfalls and uncertainties. Lessons from the crypt. *Development* 110:1001–1020, 1990.
5. Traber, P.G. and Wu, G.D. Intestinal development and differentiation. *In*: Rustgi, A.K., ed. Gastrointestinal Cancers: Biology, Diagnosis, and Therapy. Philadelphia: Lippencott-Raven Publications, 21–43, 1995.
6. Moxey, P.C. and Trier, J.S. Development of villus absorptive cells in the human fetal small intestine: A morphological and morphometric study. *Anat. Rec.* 195:463–482, 1979.
7. Menard, D. Growth-promoting factors and the development of the human gut. *In*: Lebenthal, E., ed. Human Gastrointestinal Development. New York: Raven Press, 123–150, 1989.
8. Johnson, F.P. The development of the mucous membrane of the esophagus, stomach, and small intestine in the human embryo. *Am. J. Anat.* 10:521–561, 1910.
9. Lev, R., Siegel, H.I., and Bartman, J. Histochemical studies of developing human fetal small intestine. *Histochemie* 29:103–119, 1972.
10. Verma, K.B.L. Development of mucosa of the human ileum. *J. Anat.* 128:513–521, 1979.
11. Varkonyi, T., Gergely, G., and Varro, V. The ultrastructure of the small intestinal mucosa in the developing human fetus. *Scand. J. Gastroenterol.* 9:495–500, 1974.
12. Kelley, R.O. An ultrastructural and cytochemical study of developing small intestine in man. *J. Embryol. Exp. Morph.* 29:411–430, 1973.
13. Moxey, P.C. and Trier, J.S. Specialized cell types in the human fetal small intestine. *Anat. Rec.* 191:269–286, 1978.
14. Moxey, P.C. and Trier, J.S. Endocrine cells in the human fetal small intestine. *Cell Tiss. Res.* 183:33–50, 1977.

15. Klein, R.M. and McKenzie, J.C. The role of cell renewal in the ontogeny of the intestine. I. Cell proliferation patterns in adult, fetal, and neonatal intestine. *J. Pediatr. Gastroenterol. Nutr.* 2:10–43, 1983.
16. Klein, R.M. Small intestinal proliferation during development. *In*: Lebenthal, E., ed. Human gastrointestinal development. New York: Raven Press, 367–392, 1989.
17. Klein, R.M. Cell proliferative regulation in developing small intestine. *In*: Lebenthal, E., ed. Human Gastrointestinal Development. New York: Raven Press, 393–436, 1989.
18. Sweetser, D.A., Birkenmeier, E.H., Hoppe, P.C., McKeel, D.W., and Gordon, J.I. Mechanisms underlying generation of gradients in gene expression within the intestine: an analysis using transgenic mice containing fatty acid binding protein-human growth hormone fusion genes. *Genes Develop.* 2:1318–1332, 1988.
19. Sweetser, D.A., Hauft, S.M., Hoppe, P.C., Birkenmeier, E.H., and Gordon, J.I. Transgenic mice containing intestinal fatty acid-binding protein-human growth hormone fusion genes exhibit correct regional and cell-specific expression of the reporter gene in their small intestine. *Proc. Natl. Acad. Sci. USA* 85:9611–9615, 1988.
20. Hauft, S.M., Sweetser, D.A., Rotwein, P.S., Lajara, R., Hoppe, P.C., Birkenmeier, E.H., and Gordon, J.I. A transgenic mouse model that is useful for analyzing cellular and geographic differentiation for the intestine during fetal development. *J. Biol. Chem.* 264:8419–8429, 1989.
21. Cohn, S.M., Roth, K.A., Birkenmeier, E.H., and Gordon, J.I. Temporal and spatial patterns of transgene expression in aging adult mice provide insights about the origins, organization, and differentiation of the intestinal epithelium. *Proc. Natl. Acad. Sci. USA* 88:1034–1038, 1991.
22. Kodolvsky, O. Developmental, dietary and hormonal control of intestinal disaccharidases in mammals (including man). *In*: Randle, P.J., Steiner, D.F., and Whelan, W.J., eds. Carbohydrate Metabolism and its Disorders. London: Academic Press, 481–522, 1981.
23. Galand, G. Brush-border membrane sucrase-isomaltase, maltase-glucoamylase and trehalase in mammals. *Comp. Biochem. Physiol.* 94B:1–11, 1989.
24. Henning, S.J. Ontogeny of enzymes in the small intestine. *Ann. Rev. Physiol.* 47:231–245, 1985.
25. Henning, S.J. Functional development of the gastrointestinal tract. *In*: Johnson, L.R., ed. Physiology of the Gastrointestinal Tract. 2nd ed. New York: Raven Press, 285–300, 1987.
26. Leeper, L.L. and Henning, S.J. Development and tissue distribution of sucrase-isomaltase mRNA in rats. *Am. J. Physiol.* 258:G52–G58, 1990.
27. Johnson, F.P. The development of the mucous membrane of the large intestine and vermiform process in the human embryo. *Am. J. Anat.* 14:187–233, 1913.
28. Lev, R. and Orlie, D. Histochemical and radioautographic studies of normal human fetal colon. *Histochem.* 39:301–311, 1974.
29. Lehy, T. and Cristina, M.L. Ontogeny and distribution of certain endocrine cells in the human fetal large intestine. *Cell Tiss. Res.* 203:415–426, 1979.
30. Lacroix, B., Kedinger, M., Simon-Assmann, P., Rousset, M., Zweibaum, A., and Haffen, K. Developmental pattern of brush border enzymes in the human fetal colon. Correlation with some morphogenetic events. *Early Human Develop.* 9:95–103, 1984.
31. Potter, G.D. Development of colonic function. *In*: Lebenthal, E., ed. Human Gastrointestinal Development. New York: Raven Press, 545–559, 1989.
32. Wiltz, O., O'Hara, C.J., Steele, G.D., and Mercurio, A.M. Expression of enzymatically active sucrase-isomaltase is a ubiquitous property of colon adenocarcinomas. *Gastroenterol.* 100:1266–1278, 1991.
33. Real, R.X., Xu, M., Vila, M.R., and DeBolos, C. Intestinal brush-border-associated enzymes: coordinated expression in colorectal cancer. *Int. J. Cancer* 51:173–181, 1992.
34. Czernichow, B., Simon-Assmann, P., Kedinger, M., Arnold, C., Parache, M., Marescaux, J., Zweibaum, A., and Haffen, K. Sucrase-isomaltase expression and enterocytic ultrastructure of human colorectal tumors. *Int. J. Cancer* 44:238–244, 1989.
35. Beaulieu, J.F., Weiser, M.M., Herrera, L., and Quaroni, A. Detection and characterization of sucrase-isomaltase in adult human colon and in colonic polyps. *Gastroenterol.* 98:1467–1477, 1990.
36. Tapscott, S.J., Lassar, A.B., David, R.L., and Weintraub, H. 5-Bromo-2′-deoxyuridine blocks myogenesis by extinguishing expression of myoD1. *Science* 245:532–536, 1989.
37. Gorvel, J.P., Ferrero, A., Chambraud, L., Rigal, A., Bonicel, J., and Maroux, S. Expression of sucrase-isomaltase and dipeptidylpeptidase IV in human small intestine and colon. *Gastroenterol.* 101:618–625, 1991.
38. Williams, L. and Bell, L. Asynchronous development of the rat colon. *Anat. Embryol.* 183:573–578, 1991.
39. Colony, P.C., Kois, J.M., and Peiffer, L.P. Structural and enzymatic changes during colonic maturation in the fetal and suckling rat. *Gastroenterol.* 97:338–347, 1989.

40. Sato, M. and Ahnen, D.J. Regional variability of colonocyte growth and differentiation in the rat. *Anat. Rec.* 233:409–414, 1992.
41. Foltzer-Jourdainne, C., Kedinger, M., and Raul, F. Perinatal expression of brush-border hydrolases in rat colon: hormonal and tissue regulation. *Am. J. Physiol.* 257:G496–G503, 1989.
42. Tung, J., Markowitz, A.J., Silberg, D.G., and Traber, P.G. Developmental expression of SI in transgenic mice is regulated by an evolutionarily conserved promoter. *Am. J. Physiol.* 273:G83–G92, 1997.
43. Buller, H.A., Rings E.H.H.M., Montgomery, R.K., Sybicki, M.A., and Grand, R.J. Suckling rat colon synthesizes and processes active lactase-phlorizin hydrolase immunologically identical to that from jejunum. *Pediatr. Res.* 26:232–236, 1989.
44. Birchmeier, C. and Birchmeier, W. Molecular aspects of mesenchymal-epithelial interactions. *Ann. Rev. Cell Biol.* 9:511–540, 1993.
45. Simon-Assmann, P., Kedinger, M., DeArchangelis, A., Rousseau, V., and Simo, P. Extracellular matrix components in intestinal development. *Experientia* 51:883–900, 1995.
46. Beaulieu, J.-F. Extracellular matrix components and integrins in relationship to human intestinal epithelial cell differentiation. *Progr. Histochem. Cytochem.* 31:1–78, 1997.
47. Haffen, K., Kedinger, M., and Simon-Assmann, P.M. Cell contact dependent regulation of enterocytic differentiation. *In*: Lebenthal, E., ed. Human Gastointestinal Development. New York: Raven Press, 19–40, 1989.
48. Haffen, K., Kedinger, M., and Simon-Assmann, P. Mesenchyme-dependent differentiation of epithelial progenitor cells in the gut. *J. Pediatr. Gastroenterol. Nutr.* 6:14–23, 1987.
49. Louvard, D., Kedinger, M., and Hauri, H.P. The differentiating intestinal epithelial cell:establishment and maintenance of functions through interactions between cellular structures. *Ann. Rev. Cell Biol.* 8:157–195, 1992.
50. Simon-Assmann, P., Bouziges, F., Haffen, A.K., and Kedinger, M. Epithelial-mesenchymal interactions in the production of basement membrane components in the gut. *Develop.* 102:339–347, 1988.
51. Duluc, I., Freund, J.-N., Leberquier, C., and Kedinger, M. Fetal endoderm primarily holds the temporal and positional information required for mammalian intestinal development. *J. Cell Biol.* 126: 211–221, 1994.
52. Parker, F.G., Barnes, E.N., and Kay, G.I. The pericryptal fibroblast sheath. IV. Replication, migration, and differentiation of the subepithelial fibroblasts of the crypt and villus of the rabbit jejunum. *Gastroenterol.* 67:607–621, 1974.
53. Kaestner, K.H., Silberg, D.G., Traber, P.G., and Schutz, G. The mesenchymal winged helix transcription factor Fkh6 is required for the control of gastrointestinal proliferation and differntiation. Genes *Develop.* 11:1583–1595, 1996.
54. Zang, L., Zhou, W., Veculsecu, V.E., Kern, S.E., Hruban, R.H., Hamilton, S.R., Vogelstein, B., and Kinzler, K.W. Gene expression profiles in normal and cancer cells. *Science* 276:1268–1272, 1997.
55. Markowitz, A.J., Wu, G.D., Bader, A., Cui, Z., Chen, L., and Traber, P.G. Regulation of lineage-specific transcription of the sucrase-isomaltase gene in transgenic mice and cell lines. *Am. J. Physiol.* 269:G925–G939, 1995.
56. Markowitz, A.J., Wu, G.D., Birkenmeier, E.H., and Traber, P.G. The human sucrase-isomaltase gene directs complex patterns of gene expression in transgenic mice. *Am. J. Physiol.* 265:G526–G539, 1993.
57. Traber, P.G. and Silberg, D.G. Intestine-specific gene transcription. *Ann. Rev. Physiol.* 58:275–297, 1996.
58. Wu, G.D., Chen, L., Forslund, K., and Traber, P.G. Hepatocyte nuclear factor 1α (HNF-1α) and HNF-1β regulate transcription via two elements in an intestine-specific promoter. *J. Biol. Chem.* 269:17080–17085, 1994.
59. Traber, P.G., Wu, G.D., and Wang, W. Novel DNA-binding proteins regulate intestine-specific transcription of the sucrase-isomaltase gene. *Molec. Cell Biol.* 12:3614–3627, 1992.
60. Suh, E.-R., Chen, L., Taylor, J., and Traber, P.G. A homeodomain protein related to caudal regulates intestine-specific gene transcription. *Molec. Cell Biol.* 14:7340–7351, 1994.
61. Suh, E.-R. and Traber, P.G. An intestine-specific homeobox gene regulates proliferation and differentiation. *Molec. Cell Biol.* 16:619–625, 1996.
62. Lorentz, O., Duluc, I., DeArcangelis, A., Simon-Assman, P., Kedinger, M., and Freund, J.N. Key role of the Cdx2 homeobox gene in extracellular matrix-mediated intestinal cell differentiation. *J. Cell Biol.* 139:1553–1565, 1997.
63. Duluc, I., Lorentz, O., Fritsch, C., Leberquier, C., Kedinger, M., and Freund, J.N. Changing intestinal connective tissue interactions alters homeobox gene expression in epithelial cells. *J. Cell Sci.* 110: 1317–1324, 1997.

64. Mallo, G.V., Soubeyran, P., Lissitzky, J.-C., Angre, R., Farnarier, C., Marvaldi, J., Dagorn, J.-C., and Iovanna, J.L. Expression of Cdx1 and Cdx2 homeotic genes leads to reduced malignancy in colon cancer-derived cells. *J. Biol Chem.* 273:14030–14036, 1998.
65. Subramanian, V., Meyer, B.I., and Gruss, P. Disruption of the murine homeobox gene Cdx1 affects axial skeletal identities by altering the mesodermal expression domains of Hox genes. *Cell* 83:641–653, 1995.
66. Chawengsaksophak, K., James, R., Hammond, V.E., Kontgen, F., and Beck, F. Homeosis and intestinal tumors in Cdx2 mutant mice. *Nature* 386:84–87, 1997.
67. Mallo, G.V., Rechreche, H., Frigerio, J.-M., Rocha, D., Zweibaum, A., Lacasa, M., Jordan, B.R., Busetti, N.J., Dagorn, J.-C., and Iovanna, J.L. Molecular cloning, sequencing, and expression of the mRNA encoding human CDX1 and CDX2 homeobox, down-regulation of CDX1 and CDX2 mRNA expression during colorectal carcinogenesis. *Int. J. Cancer* 74:35–44, 1997.
68. Ee, H.C., Erler, T., Bhathal, P.S., Young, G.P., and James, R.J. Cdx-2 homeodomain protein expression in human and rat colorectal adenoma and carcinoma. *Am. J. Pathol.* 147:586–592, 1995.
69. Wicking, C., Simms, L.A., Evans, T., Walsh, M., Chawengasksophak, K., Beck, F., Chenevix-Trench, G., Young, J., Jass, J., Leggett, B., and Wainwright, B. CDX2, a human homologue of *Drosophila caudal*, is mutated in both alleles in a replication error positive colorectal cancer. *Oncogene* 17:657–659, 1998.
70. Silberg, D.G., Furth, E.E., Taylor, J.K., Schuck, T., Chiou, T., and Traber, P.G. Cdx1 protein expression in normal, metaplastic and neoplastic human alimentary tract epithelium. *Gastroenterol.* 113:478–486, 1997.

2

COLONIC CELL PROLIFERATION, DIFFERENTIATION, AND APOPTOSIS

Leonard Augenlicht, Anna Velcich, John Mariadason, Michael Bordonaro, and Barbara Heerdt

Albert Einstein Cancer Center
Montefiore Medical Center
111 East 210th St.
Bronx, New York 10467

1. INTRODUCTION

Homeostasis in the colonic epithelium is established by maintaining a balance among proliferating, differentiating, and apoptotic cells. This balance is regulated by a complex interaction of intrinsic genetic signals with external stimuli including growth and dietary factors. Moreover, the interaction of signals maintains not only the quantitative balance between dividing and terminally differentiated cells, but a spatial and temporal coordination of cell proliferation, differentiation, and apoptosis within the crypt architecture as well.

Identification of structural alterations in genes associated with development of colon tumors has provided important insight into loci that regulate pathways which influence proliferation, differentiation, and apoptosis.[1] These include well studied loci such as *APC*, p53, ras, and myc, although the total number of genetic alterations in tumors are in the hundreds to thousands.[2,3] Investigations of growth factors and nutritional factors which can modulate the transformed cell phenotype have demonstrated that despite this large number of structural gene alterations which underlie tumor formation, external signals are capable of overriding biochemical pathways perturbed by genetic alterations.[4] Thus, a complex interplay among metabolic pathways has evolved that is remarkably efficient in maintaining cell homeostasis and spatial organization in the colon. This efficiency is evidenced by the fact that in the general population, colonic tumors rarely arise, and when they do, they are usually solitary, and represent a failure of the control network in a single cell of a single crypt among more than a trillion cell divisions integrated over many decades of life. We are beginning to understand individual pathways which regulate the processes of cell

Colon Cancer Prevention, edited under the auspices of AICR
Kluwer Academic / Plenum Publishers, New York, 1999.

division, differentiation, and apoptosis in colonic epithelial cells, but we know less regarding mechanisms which coordinate these processes to maintain homeostasis.

This chapter will discuss the role of short chain fatty acids in the coordination of proliferation, differentiation, and apoptosis in the colonic mucosa. Short chain fatty acids (SCFAs) are present at up to 200 mM in the lumen of the colon, produced by fermentation of fiber,[5] and they provide the principal energy source for colonic epithelial cells.[6,7] SCFAs are important in the development of the ruminant mucosa, and can induce differentiation of colonic epithelial cells *in vitro*[8,9] (and references therein) and *in vivo*.[10] We will concentrate on cell cycle arrest and apoptosis triggered by SCFAs, focusing on the role of mitochondrial function and membrane potential in commitment to and to coordinate regulation of proliferation and apoptosis; the involvement of the *B*-catenin-Tcf pathway; and how SCFAs may also play a role in coordinating lineage specific differentiation of colonic epithelial cells.

2. COORDINATION OF PROLIFERATION AND APOPTOTIC PATHWAYS IN THE COLON IN INITIATION AND PROGRESSION OF COLON CANCER

There is a spatial organization of proliferation, differentiation, and apoptosis of epithelial cells in the colon. Proliferating cells are generally confined to the lower two-thirds of the colonic crypt. As cells move upward, they undergo differentiation along multiple lineages, the most prominent being the absorptive and the secretory, or goblet cell, lineages. Cells also undergo apoptosis after they leave the proliferative compartment. However, only a few percent of colonic epithelial cells can be indentified in the terminal stages of apoptosis, utilizing the TUNEL assay.[11] This is in agreement with previous reports of low levels of apoptosis in the rat and mouse colonic mucosa.[12–14] The true number of cells which are in a cell death pathway could be larger for a number of reasons: cells in such pathways may undergo a loss of adhesiveness and be shed before exhibiting the terminal stages of apoptosis detected by TUNEL; apoptotic cells may be phagocytosed and lost without being detected by the TUNEL assay; the kinetics of completing the pathway, and thus being detected by TUNEL or other related methods may be longer than the average time it takes cells to migrate up the crypt and be lost into the lumen. Therefore, while the true number of cells which enter a cell death pathway is not clear, cells are lost from the colonic mucosa either through readily recognizable apoptosis or cell shedding, and this must balance the number of cells produced in the proliferative compartment.

In the Apc1638 mouse, the initiation of colon tumorigenesis is due to the inheritance of a mutant allele of the Apc gene which was introduced by targeted mutation into the germline.[15] This mutation produces a stop codon, which therefore gives rise to a truncated protein. Tumors arise when the wild-type allele of the Apc gene in Apc1638$^{+/-}$ mice is inactivated, most often by deletion of the entire chromosome.[16] This happens early in tumorigenesis, and is considered the initiating event. This genetic model of tumor initiation mimics the formation of colon tumors in patients with familial adenomatous polyposis (FAP), who also inherit a mutant allele of the APC gene.

In the Apc1638$^{+/-}$ mice, initiation of tumorigenesis in the intestinal mucosa is first manifest as aberrant crypt foci (ACF), morphologically abnormal crypts which form in patients at risk for tumor development and in rodents treated with colon specific carcinogens.[11] We have detected these ACF as early as 5 weeks after birth of Apc1638$^{+/-}$ mice, with substantial numbers seen at 9 weeks in the colonic mucosa. The number of ACF continues to increase over 4 fold by 22–26 weeks of age in the Apc1638 mice, but their size (i.e. number of crypts per ACF) does not increase. During this period of 9 to 26 weeks, there are no detectable changes in the number of proliferating or apoptotic cells in the gastrointestinal mucosa. Therefore, development of initiated foci in the colonic mucosa does not appear to be due to an alteration in the balance of apoptotic to proliferating cells.[11] However, the number of tumors which appear in the Apc1638 mice is increased in the mice fed a high fat diet, and this appears to be due to an effect on progression of tumors.[17] We have recently found that the high fat diet is effective in increasing cell proliferation, and substantially decreasing the ratio of apoptotic to proliferating cells in the duodenum of these mice, the principal site of tumor formation in the GI tract in this model (Augenlicht, unpublished). Thus, while initiation of tumorigenesis by an Apc mutation does not involve shifts in the ratio of apoptotic to proliferating cells, accelerated progression of initiated lesions is linked to an increase in cell proliferation in the mucosa leading to a loss of tissue homeostasis.

3. SIGNALS WHICH COORDINATE PROLIFERATION AND APOPTOSIS

The short chain fatty acid butyrate stimulates colonic carcinoma cells to enter a pathway of cell cycle arrest and apoptosis.[18,19] One of the earliest events in this pathway is the induction of the cyclin dependent kinase inhibitor, p21$^{waf1/cip1}$, which leads to a cell cycle arrest in G_0/G_1 beginning at approximately 12 hours, followed closely by a dissipation of the mitochondrial membrane potential, activation of caspase-3 and finally DNA cleavage, a marker of the end-stage of apoptosis.[20,21]

To gain insight into the mechanisms which regulate entry of cells into this pathway, we utilized scanning and image processing to build a data base on expression of each of 4000 cDNA sequences in colon carcinoma cell lines stimulated to enter this pathway by the SCFA butyrate and in colon tumors and flat mucosa.[22,23] An interesting subset of sequences was identified which were down-regulated in colon tumors compared to the normal colonic mucosa, but which returned to more normal levels when colon carcinoma cell lines were induced with butyrate.[22,23] These sequences were encoded by the mitochondrial genome.[9] Moreover, there is a tight link between the stimulation of this pathway which results in apoptosis by different agents, and upregulation of mitochondrial gene expression.[19] Non-efficiently metabolized SCFAs neither stimulate mitochondrial gene expression nor apoptosis, while dimethylsulfoxide and dimethylformamide stimulate both responses. In addition, forskolin, an inducer of MUC2 expression (a marker of the goblet cell lineage, see below), also does not stimulate mitochondrial gene expression or apoptosis.

This correlation between the ability to stimulate mitochondrial gene expression and the ability of agents to induce entry of cells into a pathway of cell cycle arrest and apoptosis was of interest since mitochondria are both structurally and functionally abnormal in colonic tumor cells.[24,25] In addition, a role for the mitochondria in the initiation and activation of apoptotic cascades has recently been demonstrated. Release of mitochondrial proteins, including cytochrome C is part of a mechanism involved in activation of caspases and triggering downstream events of apoptosis.[26] Therefore, to investigate further this link between mitochondria and the entry of cells into this pathway, we used a series of compounds which inhibit different complexes involved in mitochondrial electron transport, or in the maintenance of a mitochondrial membrane potential. We found when cells were treated with these compounds coincident with exposure to butyrate, or prior to induced caspase-3 activation, each inhibitor of mitochondrial function inhibited the DNA fragmentation characteristic of apoptosis that was induced by butyrate. However, after butyrate mediated caspase-3 activation, the cells became refractory to the mitochondrial inhibitors. Thus, once the cells have progressed through SCFA mediated cell cycle arrest and caspase activation, mitochondrial activities are dispensable and cells complete the final stages of apoptosis in the absence of electron transport or presence of a mitochondrial membrane potential. Conversely, these mitochondrial functions are necessary for the commitment of cells to this pathway.[20,21]

In these experiments, cells are grown under conditions which minmize depletion of cellular ATP levels. We therefore postulated that the maintenance of a mitochondrial membrane potential was the key to entry into this pathway, and the effects of inhibitors of electron transport were due to the necessity for electron transport in maintenance of a membrane potential, rather than ATP generation. We subsequently showed that the mitochondrial membrane potential was in fact dissipated just following cell cycle arrest.[21] Moreover, collapsing the membrane potential with valinomycin, which prevents its dissipation, inhibits the down stream events of caspase activation and DNA fragmentation. However, induction of $p21^{waf1/cip1}$ and cell cycle arrest are temporally upstream of the dissipation of the mitochondrial membrane potential, and it was therefore surprising that the collapse of the membrane potential also prevented the induction of $p21^{waf1/cip1}$ and cell cycle arrest. This demonstrates that the mitochondrial membrane potential may play a critical role in coordinating pathways which balance apoptosis and cell proliferation in the colon.[21]

4. DO PATHWAYS INDUCED BY SCFAs INTERACT WITH PATHWAYS INDUCED BY APC?

Mutations of the Apc gene initiate colon cell transformation, and it is therefore likely that the wild-type gene is normally involved in pathways associated with colonic cell differentiation and/or cell cycle regulation and apoptosis. Since this overlaps phenotypically with the effects of short chain fatty acids on colonic epithelial cells, which are physiological inducers of terminal differentiation in the colon, it is important to understand if and how these external signals interact mechanistically with Apc expression in regulating cell response.

Wild-type APC protein physically interacts with *B*-catenin and targets it for degradation.[27,28] However, mutant APC cannot form such a complex. Therefore, in cells harboring only mutant APC, *B*-catenin levels rise, as does formation of a *B*-catenin-Tcf complex which migrates to the nucleus and functions as a modulator of gene expression.[29] Thus, in the presence of mutant APC, this signalling pathway is constitutively activated, and one gene target is c-myc, which may be up-regulated, leading to tumor development and progression.[30] Butyrate treatment of colonic carcinoma cells, which stimulates an apoptotic pathway, causes up-regulation of the *B*-catenin-Tcf pathway as does mutant APC (Bordonaro and Augenlicht, unpublished). This is consistent with the finding that in *Drosophila*, mutant APC constitutively activates the pathway, and induces apoptosis in the retina.[31] Thus, up-regulation of the pathway is associated with apoptosis. Moreover, in the presence of expression of wild-type APC, which down regulates the the *B*-catenin-Tcf pathway, butyrate can induce the same level of up-regulation (Bordonaro and Augenlicht). Finally, this up-regulation by butyrate again requires the presence of a mitochondrial membrane potential, while the down-regulation induced by mutant APC is not affected by the absence of a mitochondrial membrane potential (ibid).

Therefore, it is likely that butyrate and APC affect the *B*-catenin-Tcf pathway by different mechanisms. Since butyrate does not affect *B*-catenin levels (ibid), one postulate under investigation is that butyrate acts through its ability to function as an inhibitor of histone deacetylase.

5. DIFFERENTIATION IN THE COLONIC MUCOSA

The pathway induced by SCFAs results in differentiation of colonic epithelial cells along the absorptive cell lineage, characterized by expression of alkaline phosphatase. However, the SCFA butyrate does not induce expression of the MUC2 gene.[32] MUC2 encodes the peptide backbone of the major colonic mucin, synthesized and secreted by secretory, or goblet cells. Defects in differentiation along this lineage appear to play an important role in tumorigenesis. First, aberrant crypt foci, which are early lesions in the development of colon cancer in humans, in rodents treated with colon carcinogens, and in the genetically initiated APC1638$^{+/-}$ mouse, exhibit a depletion of the goblet cell lineage.[11] Loss of this lineage and loss of mucin secretion may play a role in progression of these early lesions. Second, mucinous colon tumors, a class of tumors which have a particularly poor prognosis, have a deregulation of expression of the MUC2 gene.[33] Hence, these tumors constitutively over-express the gene, and synthesize and secrete large amounts of mucin. The link between this over-production of mucin and the aggressiveness of the tumors is not understood.

We have recently found that butyrate shuts-off MUC2 expression in cells in which it had been induced by forskolin or TPA, via PKA, and PKC dependent pathways, respectively. This inhibition of MUC2 expression by butyrate is also seen in a unique clone of HT29 colonic carcinoma cells, clone 16E cells which spontaneously express the MUC2 gene and synthesize and secrete mucin upon becoming confluent (Velcich and Augenlicht, unpublished).

Thus, the SCFA butyrate may provide a toggle switch which on the one hand triggers differentiation along the absorptive cell lineage, but which shuts off a gene fundamental to the goblet cell lineage. The mechanism by which this takes place is not known, nor is it clear if SCFAs can play this role *in vivo*. However, the potential exists that mechanisms have evolved which utilize external signals present in the colon (dietary factors, growth factors secreted by cells) to insure coordination in production of cell types necessary for colonic homeostasis.

6. QUESTIONS TO BE RESOLVED

If short chain fatty acids indeed play a pivotal role in coordinating lineage specific differentiation, proliferation, and apoptosis of cells in the colonic epithelium, it may be through various mechanisms. First, we have shown that the SCFA butyrate can up-regulate signalling through *B*-catenin-Tcf, an important mediator of effects of the APC gene. However, the up-regulation is independent of APC, and the extent to which butyrate and APC overlap in their effects on gene expression through this pathway is not clear. For example, the c-myc gene is regulated by Tcf, and increased amounts of *B*-catenin due to a mutation in APC, which consequently up-regulates Tcf signalling and expression of c-myc, have been postulated to drive tumorigenesis.[30] On the other hand, butyrate also up-regulates Tcf activity, but butyrate treatment of colonic epithelial cells results in cell cycle arrest and apoptosis, and is associated with down-regulation of c-myc expression.[34–36]

Effects of butyrate on *B*-catenin-Tcf signalling in colonic epithelial cells are relatively late events, not seen until approximately 8 hours after exposure. Therefore, earlier effects of butyrate may be brought about by other mechanisms. For example, the butyrate induction of p21$^{waf1/cip1}$ is dependent upon Sp1 sites in the p21 promotor,[37] and accessibility of these sites could involve the action of butyrate as an inhibitor of histone deacetylase.[38] In this regard, trichostatin-A, which is also an inhibitor of histone deacetylase, can trigger many of the same responses in colonic epithelial cells as does butyrate. However, the effects of both butyrate and trichostatin on colonic epithelial cells are eliminated by collapse of the mitochondrial membrane potential (Mariadason, Heerdt, and Augenlicht, unpublished). Therefore, a series of questions can be raised. Is a mitochondrial function the most up-stream event in regulating pathways of differentiation, proliferation, and/or apoptosis in colonic epithelial cells? If so, is the mitochondrial membrane potential the key to this regulation, perhaps by modulating the ability of the mitochondrial membrane to function as a docking site for various effector proteins. This could determine availability and interaction of many regulatory proteins, such as members of the Bcl-2 and Bax family, and raf1, each of which associates with the mitochondrial membrane. And finally, there are many growth factors, hormones, dietary factors, and genetic signals which can modulate colonic cell differentiation, proliferation, and apoptosis of colonic epithelial cells. If mitochondrial function, and perhaps membrane potential, play a fundamental role in coordination of these cell responses and tissue homeostasis, is mitochondrial function the integrator of multiple signals which affect these cell responses, and hence an important determinant of relative risk for the development of colonic cancer?

REFERENCES

1. Kinzler, K.W. and Vogelstein, B. Lessons from hereditary colorectal cancer. *Cell* 87:159, 1996.
2. Vogelstein, B., Fearon, E.R., Kern, S.E., Hamilton, S.R., Preisinger, A.C., Nakamura, Y., and White, R. Allelotype of colorectal carcinomas. *Science* 244:207, 1989.
3. Augenlicht, L.H., Wadler, S., Corner, G., Richards, C., Ryan, L., Multani, A.S., Pathak, S., Benson, A., Haller, D., and Heerdt, B.G. Low-level c-myc amplification in human colonic carcinoma cell lines and tumors: a frequent, p53-independent mutation associated with improved outcome in a randomized multi-institutional trial. *Cancer Res.* 57:1769, 1997.
4. Augenlicht, L.H., Velcich, A., and Heerdt, B.G. Short chain fatty acids and molecular and cellular mechanisms of colonic cell differentiation and transformation, *In*: Diet and Cancer: Molecular Mechanisms of Interactions (eds. Amer. Inst. Cancer Res.) Plenum, NY 1995.
5. Cummings, J.H. Fermentation in the human large intestine: evidence and implications for health. *Lancet* ••:1206, 1983.
6. Roediger, W.E. Role of anaerobic bacteria in the metabolic welfare of the colonic mucosa in man. *Gut* 21:793, 1980.
7. Roediger, W.E.W. Bacterial short-chain fatty acids and mucosal diseases of the colon. *Br. J. Surg.* 75:346, 1988.

8. Kim, Y.S., Tsao, D., Siddiqui, B., Whitehead, J.S., Arnstein, P., Bennett, J., and Hicks, J. Effects of sodium butyrate and dimethylsulfoxide on biochemical properties of human colon cancer cells. *Cancer* 45:1185, 1980.
9. Heerdt, B.G., Halsey, H.K., Lipkin, M., and Augenlicht, L.H. Expression of mitochondrial cytochrome c oxidase in human colonic cell differentiation, transformation, and risk for colonic cancer. *Cancer Res.* 50:1596, 1990.
10. Harig, J.M., Soergel, K.H., Komorowski, R.A., and Wood, C.M. Treatment of diversion colitis with short-chain-fatty acid irrigation. *N. Engl. J. Med.* 320:23, 1989.
11. Augenlicht, L.H., Heerdt, B.G., Anthony, G.M., Church, T.L., Edelmann, W., Kucherlapati, R., Hudson, L., and Pretlow, T.P. Aberrant crypt formation, proliferation, and apoptosis in the flat intestinal mucosa of Apc1638 mice which inherit a mutant Apc allele. Submitted, 1998.
12. Risio, M., Lipkin, M., Newmark, H., Yang, K., Rossini, F.P., Steele, V.E., Boone, C.W., and Kelloff, G.J. Apoptosis, cell replication, and western-style diet-induced tumorigenesis in the mouse colon. *Cancer Res.* 56:4910, 1996.
13. Chang, W.-C.L., Chapkin, R.S., and Lupton, J.R. Predictive value of proliferation, differentiation, and apoptosis as intermdiate markers for colon tumorigenesis. *Carcinogenesis* 18:721, 1997.
14. Samaha, H.S., Kelloff, G.J., Steele, V., Rao, C.V., and Reddy, B.S. Modulation of apoptosis by sulindac, curcumin, phenylethyl-3-methylcaffeate, and 6-phenylhexyl isothiocyanate: apoptotic index as a biomarker in colon cancer chemoprevention and promotion. *Cancer Res.* 57:1301, 1997.
15. Fodde, R., Edelmann, W., Yang, K., van Leeuwen, C., Carlson, C., Renault, B., Breukel, C., Alt, E., Lipkin, M., Khan, P.M., and Kucherlapati, R. A targeted chain-termination mutation in the mouse Apc gene results in multiple intestinal tumors. *Proc. Natl. Acad. Sci. USA* 91:8969, 1994.
16. Yang, K., Edelmann, W., Fan, K., Lau, K., Kolli, V.R., Fodde, R., Khan, P.M., Kucherlapati, R., and Lipkin, M. A mouse model of human familial adenomatous polyposis. *J. Exp. Zool.* 277:245, 1997.
17. Lipkin, M., Newmark, H., Yang, K., Fan, K., Kucherlapati, R., Edelmann, W., and Leung, D. Modulation of colonic lesions in Apc1638 mice by a Western-style diet. Submitted, 1994.
18. Heerdt, B.G., Houston, M.A., and Augenlicht, L.H. Potentiation by specific short-chain fatty acids of differentiation and apoptosis in human colonic carcinoma cell lines. *Cancer Res.* 54:3288, 1994.
19. Heerdt, B.G., Houston, M.A., Rediske, J.J., and Augenlicht, L.H. Steady-state levels of mitochondrial messenger RNA species characterize a predominant pathway culminating in apoptosis and shedding of HT29 human colonic carcinoma cells. *Cell Growth Differ.* 7:101, 1996.
20. Heerdt, B.G., Houston, M.A., and Augenlicht, L.H. Short chain fatty acid-initiated cell cycle arrest and apoptosis of colonic epithelial cells is linked to mitochondrial function. *Cell Growth Differ.* 8:523, 1997.
21. Heerdt, B.G., Houston, M.A., Anthony, G.M., and Augenlicht, L.H. Mitochondrial membrane potential in the coordination of p53-independent proliferation and apoptosis pathways in human colonic carcinoma cells. *Cancer Res.* 58:2869, 1998.
22. Augenlicht, L.H., Wahrman, M.Z., Halsey, H., Anderson, L., Taylor, J., and Lipkin, M. Expression of cloned sequences in biopsies of human colonic tissue and in colonic carcinoma cells induced to differentiate *in vitro*. *Cancer Res.* 47:6017, 1987.
23. Augenlicht, L.H., Taylor, J., Anderson, L., and Lipkin, M. Patterns of gene expression that characterize the colonic mucosa in patients at genetic risk for colonic cancer. *Proc. Natl. Acad. Sci. USA* 88:3286, 1991.
24. Summerhayes, I.C., Lampidis, T.J., Bernal, S.D., Nadakavukaren, J.J., Nadakaukaren, K.K., Shepherd, E.L., and Chen, L.B. Unusual retention of rhodamine 123 by mitochondria in muscle and carcinoma cells. *Proc. Natl. Acad. Sci. USA* 79:5292, 1982.
25. Modica-Napolitano, J.S., Steele, G.D., and Chen, L.B. Aberrant mitochondria in two human colon carcinoma cell lines. *Cancer Res.* 49:3369, 1989.
26. Reed, J.C. Cytochrome c: Can't live with it-Can't live without it. *Cell* 91:559, 1997.
27. Rubinfeld, B., Souza, B., Albert, I., Muller, O., Chamberlain, S.H., Masiarz, F.R., Munemitsu, S., and Polakis, P. Association of the APC gene product with b-catenin. *Science* 262:1731, 1993.
28. Su, L.-K., Vogelstein, B., and Kinzler, K. Association of the APC tumor suppressor protein with catenins. *Science* 262:1734, 1993.
29. Korinek, V., Barker, N., Morin, P.J., van Wichen, D., de Weger, R., Kinzler, K.W., Vogelstein, B., and Clevers, H. Constitutive transcriptional activation by a b-catenin-Tcf complex in APC–/– colon carcinoma. *Science* 275:1784, 1997.
30. He, T.-C., Sparks, A.B., Rago, C., Hermeking, H., Zawel, L., da Costa, L.T., Morin, P.J., Vogelstein, B., and Kinzler, K.W. Identification of c-MYC as a target of the APC pathway. *Science* 281:1509, 1998.

31. Ahmed, Y., Hayashi, S., Levine, A., and Wieschaus, E. Regulation of Armadillo by a Drosophila APC inhibits neuronal apoptosis during retinal development. *Cell* 93:1171, 1998.
32. Velcich, A., Palumbo, L., Jarry, A., Laboisse, C., Rachevskis, J., and Augenlicht, L. Patterns of expression of lineage specific markers during the *in vitro* differentiation of HT29 colon carcinoma cells. *Cell Growth Differ.* 6:749, 1995.
33. Velcich A. and Augenlicht, L.H. Regulated expression of an intestinal mucin gene in HT29 colonic carcinoma cells. *J. Biol. Chem.* 268:13956, 1993.
34. Herold K.M. and Rothberg, P.G. Evidence for a labile intermediate in the butyrate induced reduction of the level of *c-myc* RNA in SW837 rectal carcinoma cells. *Oncogene* 3:423, 1988.
35. Taylor, C.W., Kim, Y.S., Childress-Fields, K.E., and Yeoman, L.C. Sensitivity of nuclear *c-myc* levels and induction to differentiation-inducing agents in human colon tumor cell lines. *Cancer Lett.* 62:95, 1992.
36. Barnard, J.A. and Warwick, G. Butyrate rapidly induces growth inhibition and differentiation in HT-29 cells. *Cell Growth Differ.* 4:495, 1993.
37. Nakano, K., Mizuno, T., Sowa, Y., Orita, T., Yoshino, T., Okuyama, Y., Fujita, T., Ohtani-Fujita, N., Matsukawa, Y., Tokino, T., Yamagishi, H., Oka, T., Nomura, H., and Sakai, T. Butyrate activates the waf1/cip1 gene promoter through Sp1 sites in a p53-negative human colon cancer cell line. *J. Biol. Chem.* 272:22199, 1997.
38. Medina, V., Edmonds, B., Young, G.P., James, R., Appleton, S., and Zalewski, P.D. Induction of caspase-3 protease activity and apoptosis by butyrate and trichostatin A (inhibitors of histone deacetylase): dependence on protein synthesis and synergy with a mitochondrial/cytochrome c-dependent pathway. *Cancer Res.* 57:3697, 1997.

3

DEFECTS IN THE REGULATION OF β–CATENIN IN COLORECTAL CANCER

Paul Polakis, Matt Hart, and Bonnee Rubinfeld

Onyx Pharmaceuticals
3031 Research Drive
Richmond, California 94806

ABSTRACT

The molecular events that contribute to the progression of colon cancer are beginning to unravel. An initiating and probably obligatory event is the oncogenic activation of β–catenin. This can come about by the loss of its negative regulator the adenomatous polyposis coli (APC) protein, or by mutations in the β–catenin gene that result in a more stable protein product. The interaction between APC and β–catenin, and additional proteins that affect assembly and signaling along this pathway, are discussed.

1. OVERVIEW

The progression of neoplasia in the large intestine is dependent upon a succession of genetic errors that ultimately lead to the genesis of malignant cancers. Among the earliest known mutations in this progression is the inactivation of the adenomatous polyposis coli (APC) tumor suppressor gene.[1] Mutations in APC have been detected in the earliest stages of dysplasia, referred to as aberrant crypt foci, and the vast majority all tumors that arise in the colon contain mutations in this gene.[2] The prevalence of APC mutations in colon cancer and their presence in the germline of individuals predisposed to this disease underscores the importance of this tumor suppressor in maintaining homeostatic growth control in the epithelium of the gut. In murine genetic models, animals heterozygotic for APC always develop spontaneous intestinal neoplasia[3] and the forced elimination of the remaining wildtype APC allele results in the formation of tumors within as little as three weeks time.[4] Thus APC has been named the gatekeeper of colon cancer and its mutation is regarded as essential for progression of

Colon Cancer Prevention, edited under the auspices of AICR
Kluwer Academic / Plenum Publishers, New York, 1999.

this disease.[1] The delineation of APC gene function is critical to our understanding of colon cancer and ultimately to our ability to manage this disease successfully.

The identification of the APC gene by positional cloning was a hallmark finding that set the stage for further studies that have recently allowed the development of signaling pathway models. This pathway was in part based on studies in *Xenopus* and *Drosophila* where signaling by the wnt-1 oncogene, and its *Drosophila* homolog wingless, leads to cell fate determinations that serve as outputs in establishing epistatic relationships between the participating genes.[5,6] The rationale for positioning APC in the wnt-/wingless pathway relates to its physical interaction with β–catenin, a key intermediate signaling component in the *Xenopus* and *Drosophila* model systems. In these models wnt-/wingless is proposed to activate β–catenin, or its *Drosophila* homolog armadillo, in a manner that regulates the transcription of target genes.[5,6] A productive signal involves the physical interaction of β–catenin with the high mobility group transcription factors TCF/LEF.[7]

The mechanism of wnt-1 signaling involves the post translational stabilization of β–catenin, a protein that normally exhibits rapid turnover in the cell.[8] The ability of wnt-1 to stabilize β–catenin, and the propensity for tumor formation following ectopic expression of wnt-1 in murine mammary tissue,[9] points to β–catenin as a mediator of the cancer process. In accordance with this hypothesis, it was found that the APC tumor suppressor negatively regulated β–catenin when introduced into colon cancer cells.[10] The downregulation of β–catenin by APC was abrogated by the polypeptide chain-terminating mutations that are typically identified in colon cancers. Experimentally, a strong correlation was found to exist between the position of the mutation and the loss of β–catenin regulation in the resulting APC polypeptide.[11] Thus, the upregulation of β–catenin by the oncogene wnt-1 and its downregulation by the tumor suppressor APC placed β–catenin at the nexus of the cancer signaling pathway in colonic epithelia. Moreover, another component of the wnt-1 signaling pathway, glycogen synthase kinase 3β (GSK3β), interacted with APC and phosphorylated it at serine and threonine residues that promoted its association with β–catenin.[12] This kinase was also reported to phosphorylate β–catenin at serine and threonine residues known to be involved in regulating the stability of the protein.[13]

The model implicating β–catenin in colon cancer rests largely on complimentary findings that suggest it is the stabilization of the β–catenin protein that endows it with oncogenic properties. This view is consistent with recent work describing the identification of mutations in the β–catenin gene in primary human cancers. These mutations were initially identified in melanomas and in those rare colon cancers in which APC was wildtype.[14,15] Importantly, the missense mutations affected the same serine and threonine residues known to be phosphorylated by GSK3β. Introduction of these mutations into recombinant β–catenin resulted in a protein with a much longer half-life than wildtype protein when expressed in cells that normally exhibited a rapid turnover of this protein.[14] Additional mutations that resulted in interstitial deletion of nucleotide sequence, containing the codons for these serine and threonine residues were identified.[16,17] The publication of numerous recent studies describing the prevalence of β–catenin mutations in a variety of human cancers leaves little doubt that this gene is indeed proto-oncogenic.[16,18–22]

The events that follow the oncogenic activation of β–catenin are now under intense investigation. It is likely that cancer is driven in part by the ability of β–catenin to interact with and regulate transcription by the TCF/LEF transcription factors.[23,24] A hunt for genes that are turned on by the β–catenin pathway has resulted in the identification of the well-known proto-oncogene c-myc.[25] Although certainly important, it is unlikely that c-myc is the only target of β–catenin in cancer signaling. This qualification is based on the finding that tumors originating in livers of rodents transgenically expressing c-myc also contain, at high frequency, spontaneous oncogenic mutations in β–catenin.[18] This suggests that c-myc and β–catenin operate in parallel rather than in a simple linear relationship, as the activation of myc expression by β–catenin might imply. It must also be kept in mind that β–catenin physically interacts with numerous substrates in the cell including the EGF receptor, certain tyrosine phosphatases, cytoskeletal proteins, and a regulator of the small GTPases IQGAP.[26–29] However, this does not diminish the importance of β–catenin as an oncogene, and any rational approach to the therapeutic management of colon cancer should include the consideration of this molecule.

The control of β–catenin stability in the cell has therefore been a key interest for those attempting to interfere with the cancer signal propagated along this pathway. Here I will discuss studies that have lead to the identification of some of the key components in the regulation of β–catenin and β–catenin signaling.

2. DOWNREGULATION OF β–CATENIN BY APC

The immunochemical staining of colorectal cancer cells for β–catenin revealed an excess level of this protein that was localized throughout the cell as opposed to the typical pattern of staining that is confined to areas of cell-cell contact.[10] A comparison of the human 293 cell to the colorectal cancer cell SW480 is presented, where the latter cell is wildtype for APC and the former is mutant (Fig. 1A). Introduction of full-length wildtype APC into the SW480 cell by transient transfection resulted in the disappearance of the excess β–catenin from approximately 50% of the cells (Fig. 1B). This is roughly equivalent to the transfection efficiency achieved with these cells. Moreover, when cells that expressed the ectopic wildtype APC were identified and then costained for β–catenin, a loss of β–catenin signal was invariably noted. An examination of the process by which the elimination of β–catenin was effected, revealed a posttranslational mechanism that occurred at the level of the protein. Deletional analysis of APC indicated that it was the central region of the protein that harbored the activity, and a more precise localization demonstrated a correlation between activity and the presence of the 20-amino acid repeat sequences (Fig. 1C). The loss of activity occured as the truncations approached the 3′ border of the mutation cluster region (MCR)[30] where the vast majority of the somatic APC mutations are localized. These results suggest that APC regulates the levels of β–catenin in the cell and that mutations that truncate the APC protein in cancer eliminate this activity.

3. THE RELATIONSHIP OF β–CATENIN TO GSK3β

It was apparent from studies in *Drosophila* that armadillo was acutely regulated by the GSK3β homolog zeste white 3.[31] Disruption of this gene led to high levels of cytoplasmic and nuclear armadillo staining in the cells of gastrulating animals. Moreover, expression of dominant negative GSK3β in *Xenopus* embryos phenocopied the duplication of the dorsal axis that is characteristic of wnt-1 signaling in this system. These results prompted us to evaluate the potential interaction between β–catenin,

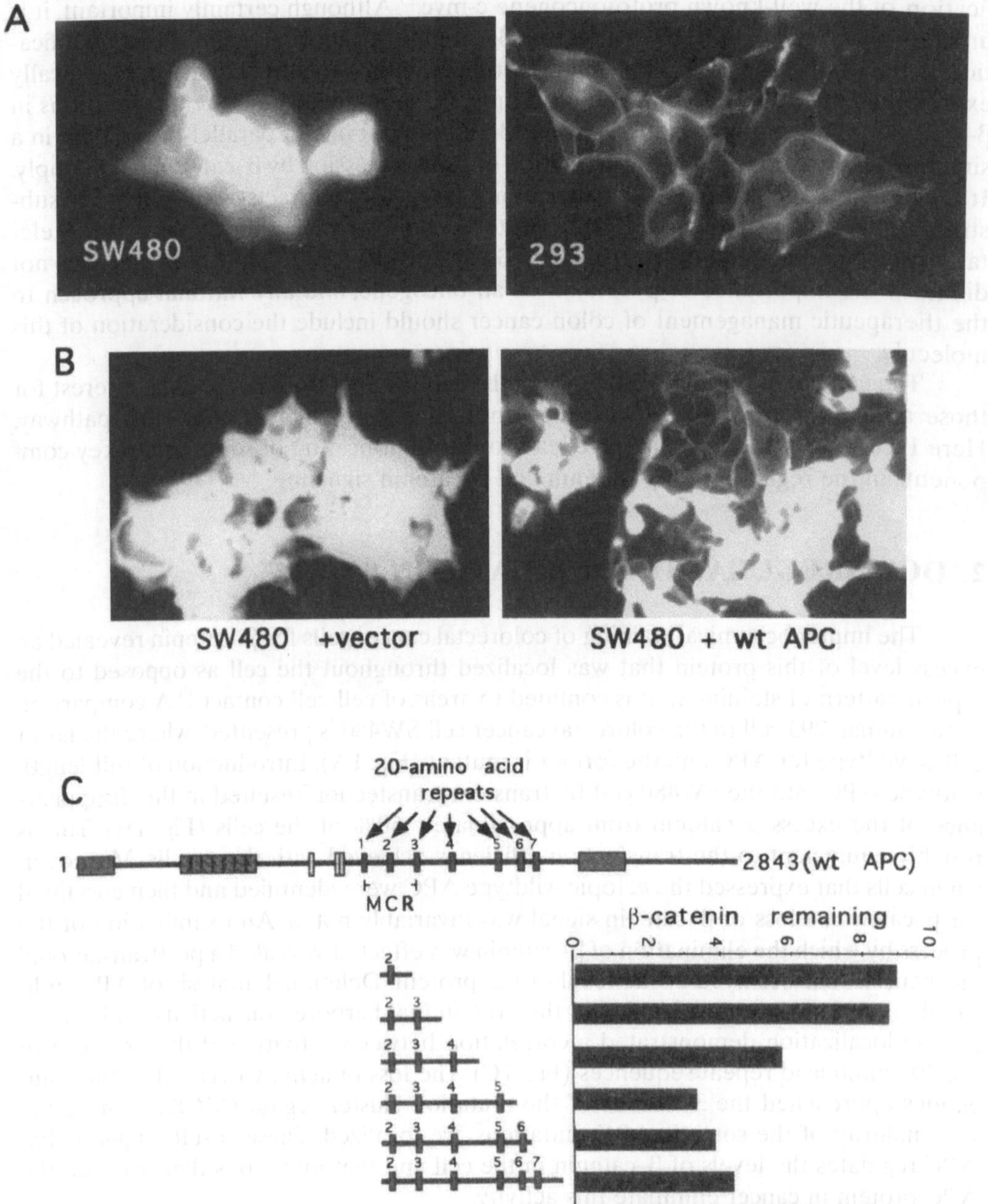

Figure 1. Downregulation of β–catenin by APC. (A) Immunofluorescent staining for β–catenin in the human kidney cell line 293 and the human colorectal cancer cell line SW480. (B) Transient transfection of SW480 cells with empty vector or a plasmid expressing wildtype APC followed by immunofluorescent staining for β–catenin. (C) Downregulation of β–catenin by various fragments of APC transfected into SW480 cells. MCR indicates the mutation cluster region and demarcates the termination of the APC polypeptide chain resulting from mutations in the corresponding region of the open reading frame of the gene.

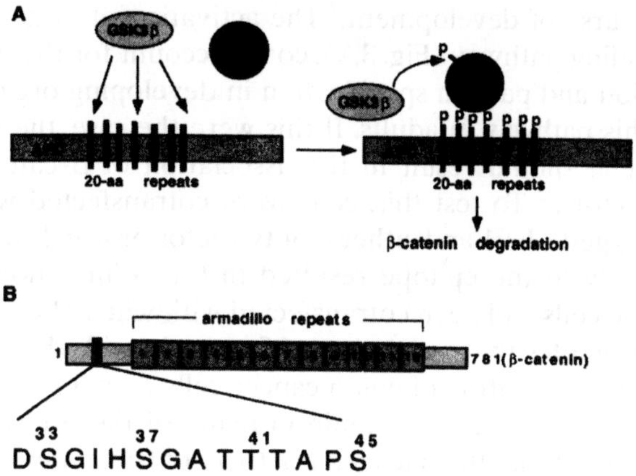

Figure 2. Phosphorylation of APC and β–catenin by GSK3β. (A) The model depicts the phosphorylation of the 20-amino acid repeat sequences in the central region of APC by GSK3β. The phosphorylation promotes the binding of β–catenin to this region of APC where β–catenin also undergoes phosphorylation by GSK3β leading to its targeted degradation. (B) The sites of phosphorylation by GSK3β in β–catenin (S33, S37, T41, and S45) are localized to an amino-terminal regulatory domain in which mutations have been identified in human tumors.

APC, and GSK3β in mammalian cells. Indeed, immunoprecipitation of GSK3β from colon cancer cells resulted in the detection of both β–catenin and APC protein in the immune complexes.[12] The association of GSK3β with APC and β–catenin was not detected in cells containing wildtype APC unless a mutant stabilized form of β–catenin was ectopically expressed. Importantly, APC served as an excellent substrate for GSK3β *in vitro* where several moles of phosphate per mole of APC were incorporated. This phosphorylation was localized to the central region of the APC polypeptide chain where the seven tandem repeats of 20 amino acids each are localized. Each of these repeat sequences bears a consensus site (SXXXS) for GSK3β phosphorylation and fragments of APC containing these sequences are capable of downregulating β–catenin when expressed in cancer cells. The phosphorylation of the central region of APC by GSK3β *in vitro* promoted its subsequent binding to β–catenin. Thus, the regulatory region of APC, that is typically truncated by cancer causing mutations, also binds β–catenin in a manner dependent upon phosphorylation by GSK3β, a kinase shown to regulate β–catenin in developmental model systems (Fig. 2A).

The finding that β–catenin was also present in a complex with GSK3β suggested that it too served as a substrate for this kinase. This was found to be the case and four sites of phosphorylation were localized to serine/threonine residues encompassed by amino-terminal amino acids 33–45.[13] Mutation of these amino acids in β–catenin greatly extended its half-life in cells containing wildtype APC and also prevented the downregulation of β–catenin by ectopically expressed APC.[14] These experiments demonstrated that GSK3β phosphorylates APC to facilitate its interaction with β–catenin, and phosphorylates β–catenin to initiate its APC-dependent turnover in the cell. The regulatory sequence in β–catenin is presented in Fig. 2B with the GSK3β phosphorylation sites indicated as S33, S37 T41, and S45. All four of these amino acids have now been found mutated in human cancers.

4. INTERACTION OF β–CATENIN WITH TRANSCRIPTION FACTORS

The discovery that β–catenin associated with the LEF/TCF transcription factors was consistent with the ability of β–catenin to effect cell fate determination during the course of development.[7] The activation of target genes, as predicted by the wnt-1 signaling pathway (Fig. 3A), could account for the complex processes driving differentiation and pattern specification in developing organisms and the oncogenic potential of this pathway in adults. If this were the case, the activation of cells by the wnt-1 oncogene should result in the association of β-catenin with the LEF/TCF transcription factors.[32] To test this, cells were cotransfected with a plasmid coding for an epitope tagged LEF1 and either empty vector or wnt-1. Immunoprecipitation of LEF-1 by antibody to the epitope resulted in the coimmunoprecipitation of β-catenin only when the cells had been cotransfected with wnt-1 (Fig. 3B). Thus, wnt-1 provides a signal that promotes the association of β-catenin with the LEF-1 transcription factor. Moreover, an examination of colon cancer cells demonstrated that LEF-1 was also constitutively associated with β-catenin in many of the transformed cells (Fig. 3C). These findings suggest that the oncogenic activity of β-catenin is, in part, driven by its interaction with the LEF/TCF transcription factors.

5. AXIN IS A SCAFFOLD PROTEIN THAT INDEPENDENTLY BINDS TO APC, β-CATENIN, AND GSK3β

The participation of GSK3β in a diverse number of cellular processes suggests that there is a mechanism by which the kinase is localized to specific substrates, thereby preventing cross talk between unrelated signaling pathways. A mechanism for this type of segregation as it relates to β-catenin signaling has been revealed by the discovery of the axin protein. The gene coding for the axin protein was originally identified by murine genetic studies and subsequently reported to inhibit wnt-1 signaling when expressed in *Xenopus* embryos.[33] Implications for axin in wnt-1 signaling prompted us to examine the relationship between axin and other wnt-1 protein components.[34] The binding of axin to APC, β-catenin, and GSK3β was assessed by testing the association of these purified recombinant proteins with axin produced by *in vitro* translation. The binding of the central fragment of APC to axin was mediated by the amino-terminal RGS domain in the axin protein (Fig. 4A). By contrast, binding to β-catenin and GSK3β was preserved in axin fragments lacking the RGS domain and was localized to amino acids 320–513 of the full-length 900-amino acid axin polypeptide. In a reciprocal deletional analysis with APC, binding of axin was mediated by a site residing between the third and fourth repeats of 20 amino acids each. However, additional binding occurred with fragments not containing this region but containing sequence carboxyl-terminal to it. Further analysis by ourselves and Birchmeier and colleagues[35] have revealed the presence of at least three independent binding sites for axin as indicated in Fig. 4B. The position of the most amino-terminal of these three sites lies very close to the boundary at which the mutations in the APC gene are predicted to truncate the protein. The mapping of these sites suggests that the binding of APC to axin is selected against in cancer.

The interaction of axin with APC at sites that correlate with clinically relevant mutations suggested that axin might participate in the downregulation of β-catenin. To test this, axin was overexpressed in SW480 colon cancer cells and the levels of β-catenin were visualized by immunofluorescent microscopy.[34] Expression of wildtype axin eliminated the pool of cytoplasmic and nuclear β-catenin present in these cells. Surprisingly, a mutant form of axin lacking the RGS domain, and therefore the APC binding site, also downregulated β-catenin. Further analysis of the downregulation of β-catenin

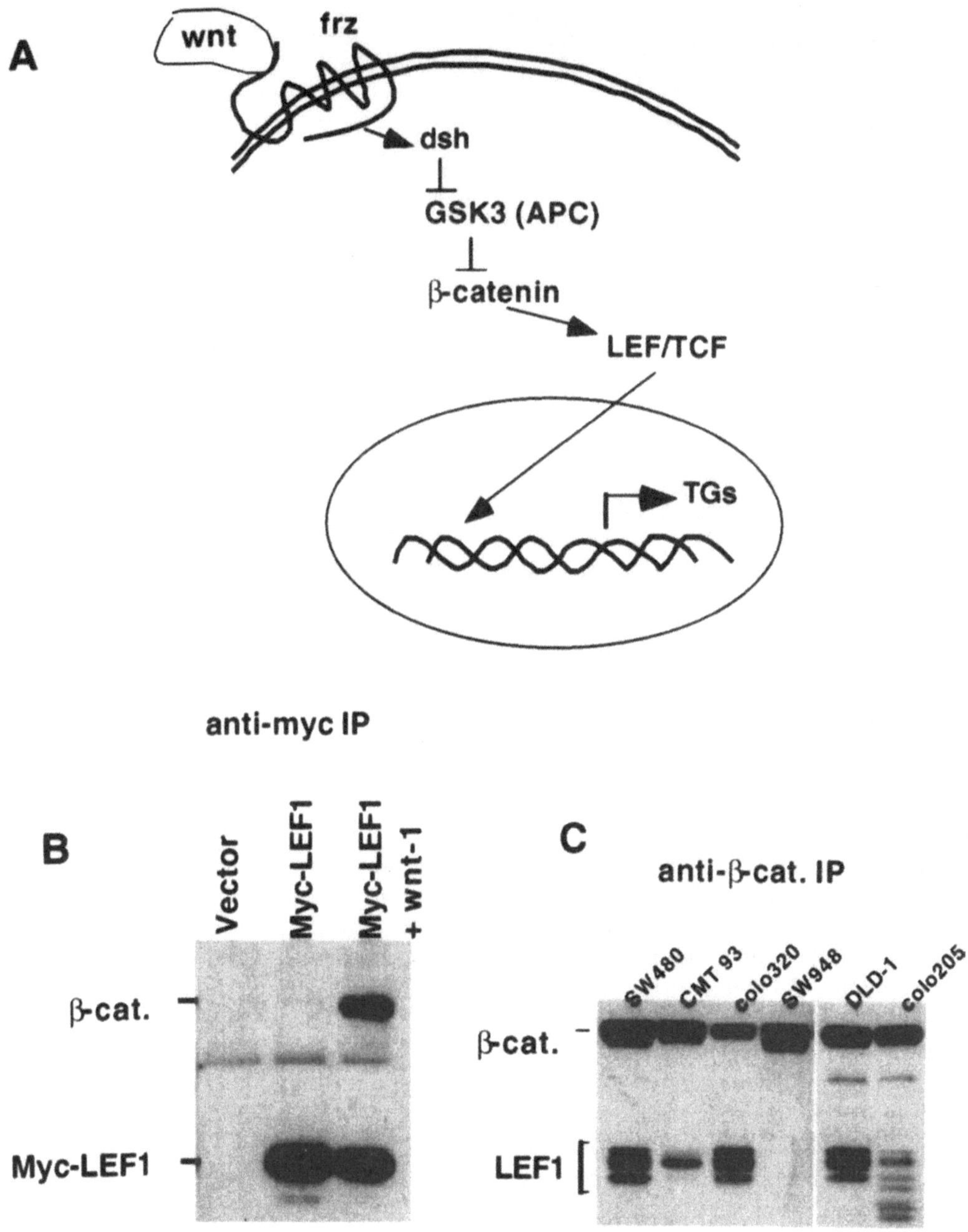

Figure 3. Association of β–catenin with the LEF transcription factor. (A) The wnt-1 pathway is activated by the binding of wnt-1 ligand to the receptor frizzled (frz) that signals to the disheveled protein resulting in the inactivation of GSK3β. GSK3β is required for downregulation of β–catenin, and its inhibition results in the accumulation of β–catenin that ultimately binds to the LEF/TCF transcription factors resulting in the activation of target genes (TGs) in the nucleus. (B) Cotransfection of human 293 cells with myc-epitope tagged LEF1 and wnt-1 resulted in the coimmunoprecipitation of LEF1 and β–catenin by antibody to the myc epitope. (C) Antibodies specific to β–catenin coimmunoprecipitated LEF1 from the indicated colon cancer cell lines.

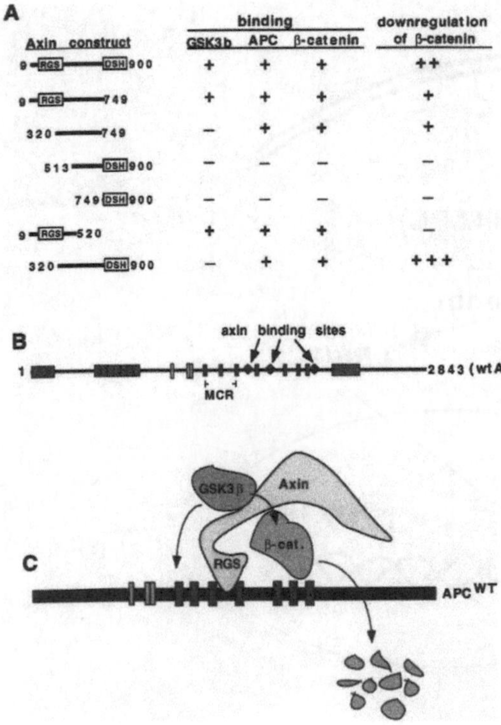

Figure 4. Interaction of APC with axin. (A) The indicated fragments of axin were tested for binding to GSK3β, APC, and β–catenin *in vitro* and for the downregulation of β–catenin in colon cancer cells. (B) Localization of axin binding sites in APC. (C) Proposed mechanism for the involvement of axin in the downregulation of β–catenin. Axin binds GSK3β, APC, and β–catenin at independent sites promoting the phosphorylation of the latter two proteins and thereby targeting β–catenin for degradation.

by pulse-chase analysis demonstrated that axin expression enhanced the turnover of wildtype β–catenin but not that of the serine37alanine mutant.[34] The rate of turnover of β–catenin was even greater on expression of the axin deletion mutant lacking the RGS domain. The interpretation of these results must accommodate the fact that axin downregulates β–catenin in cells that lack functional APC and that axin performs this activity despite the absence of an APC binding site. A model that is consistent with these observations proposes that APC act as an activator of axin (Fig. 4C). In cells lacking APC the latent activity of axin to downregulate β–catenin is not engaged and downregulation proceeds at a slow rate. It is the role of APC to derepress axin by interacting with its RGS domain. Removal of the RGS domain from axin, which houses the APC binding site, artificially derepresses axin activity, and explains the increased rate of β–catenin downregulation observed with this mutant. The ability of ectopically expressed wildtype axin to downregulate β–catenin appears strong, but this is due to mass action provided by the high levels of protein achieved by overexpression—perhaps as high as 100-fold over endogenous axin. We propose that axin binding sites are critical to the ability of APC to regulate β–catenin. These sites are typically deleted from APC in cancer suggesting they are selected against in cancer progression. Indeed, families in which the germline mutation in APC occurs at codon 1597, thereby eliminating only two of the three axin binding sites from the expressed APC mutant, represent a mild phenotype.[36] These patients develop far fewer polyps than full blown familial adenomatous polyposis patients, and they do not develop malignant cancers until a much later age.

6. CONCLUSIONS

The β–catenin protein is a central obligatory player in the progression of colon cancer. It is regulated by a complex process that involves its phosphorylation at specific serine and threonine residues. These phosphorylations likely set up β–catenin for targeted degradation by the proteosome and thereby control its levels in the cell. So far, genetic mutations that affect either APC or β–catenin can account for the defects in this pathway in about 90% of all colon cancers. It is possible that genetic or regulatory defects that are exclusive of either APC or β–catenin could account for an additional small percentage of cancers driven by this pathway. Importantly, the therapeutic manipulation of the β–catenin pathway appears to be a promising avenue for the clinical management of colon cancer.

REFERENCES

1. Kinzler, K.W. and Vogelstein, B. Lessons from hereditary colorectal cancer. *Cell* 87:159–170, 1996.
2. Smith, A.J., Stern, H.S., Penner, M., Kazy, H., Mitri, A., Bapat, B.V., and Gallinger, S. Somatic APC and K-ras codon 12 mutations in aberrant crypt foci from human colons. *Cancer Res.* 54:5527–5530, 1994.
3. Su, L.-K., Kinzler, K.W., Vogelstein, B., Presinger, A.C., Moser, A.P., Luongo, C., Gould, K.A., and Dove, W.F. Multiple intestinal neolasia caused by a mutation in the murine homolog of the APC gene. *Science* 256:668–670, 1992.
4. Shibata, H., Toyama, K., Shioya, H., Ito, M., Hirota, M., Hasegawa, S., Matsumoto, H., Takano, H., Akiyama, T., Toyoshima, K., Kanamaru, R., Kanegae, Y., Saito, I., Nakamura, Y., Shiba, K., and Noda, T. Rapid colorectal adenoma formation initiated by conditional targeting of the Apc gene. *Science* 278:120–123, 1997.
5. Peifer, M. Cell adhesion and signal transduction: The armadillo connection. *Trends Cell. Biol.* 5:224–229, 1995.
6. Gumbiner, B.M. Signal transduction of beta-catenin. *Curr. Opin. Cell Biol.* 7:634–640, 1995.
7. Miller, J.R. and Moon, R.T. Signal transduction through b-catenin and specification of cell fate during embryogenesis. *Genes Dev.* 10:2527–2539, 1996.
8. Hinck, L., Nelson, W.J., and Papkoff, J. Wnt-1 modulates cell-cell adhesion in mammalian cells by stabilizing b-catenin binding to the cell adhesion protein cadherin. *J. Cell Biol.* 124:729–741, 1994.
9. Tsukamoto, A.S., Grosschedl, R., Guzman, R.C., Parslow, T., and Varmus, H.E. Expression of the int-1 gene in transgenic mice is associated with mammary gland hyperplasia and adenocarcinomas in male and female mice. *Cell* 55:619–625, 1988.
10. Munemitsu, S., Albert, I., Souza, B., Rubinfeld, B., and Polakis, P. Regulation of intracellular beta-catenin levels by the adenomatous polyposis coli (APC) tumor-suppressor protein *Proc. Natl. Acad. Sci. USA* 92:3046–3050, 1995.
11. Rubinfeld, B., Albert, I., Porfiri, E., Munemitsu, S., and Polakis, P. Loss of beta-catenin regulation by the APC tumor suppressor protein correlates with loss of structure due to common somatic mutations of the gene. *Cancer Res.* 57:4624–4630, 1997.
12. Rubinfeld, B., Albert, I., Porfiri, E., Fiol, C., Munemitsu, S., and Polakis, P. Binding of GSK3beta to the APC-beta-catenin complex and regulation of complex assembly. *Science* 272:1023–1026, 1996.
13. Yost, C., Torres, M., Miller, J.R., Huang, E., Kimelman, D., and Moon, R.T. The axis-inducing activity, stability, and subcellular distribution of beta-catenin is regulated in *Xenopus* embryos by glycogen synthase kinase 3. *Genes Dev.* 10:1443–1454, 1996.
14. Rubinfeld, B., Robbins, P., El-Gamil, M., Albert, I., Porfiri, E., and Polakis, P. Stabilization of beta-catenin by genetic defects in melanoma cell lines. [see comments] *Science* 275:1790–1792, 1997.
15. Morin, P.J., Sparks, A.B., Korinek, V., Barker, N., Clevers, H., Vogelstein, B., and Kinzler, K.W. Activation of beta-catenin-Tcf signaling in colon cancer by mutations in beta-catenin or APC. *Science* 275:1787–1790, 1997.

16. Miyoshi, Y., Iwao, K., Nagasawa, Y., Aihara, T., Sasaki, Y., Imaoka, S., Murata, M., Shimano, T., and Nakamura, Y. Activation of the beta-catenin gene in primary hepatocellular carcinomas by somatic alterations involving exon 3. *Cancer Res.* 58:2524–2527, 1998.
17. Iwao, K., Nakamori, S., Kameyama, M., Imaoka, S., Kinoshita, M., Fukui, T., Ishiguro, S., Nakamura, Y., and Miyoshi, Y. Activation of the beta-catenin gene by interstitial deletions involving exon 3 in primary colorectal carcinomas without adenomatous polyposis coli mutations. *Cancer Res.* 58:1021–1026, 1998.
18. de La Coste, A., Romagnolo, B., Billuart, P., Renard, C.A., Buendia, M.A., Soubrane, O., Fabre, M., Chelly, J., Beldjord, C., Kahn, A., and Perret, C. Somatic mutations of the beta-catenin gene are frequent in mouse and human hepatocellular carcinomas. *Proc. Natl. Acad. Sci. USA* 95:8847–8851, 1998.
19. Fukuchi, T., Sakamoto, M., Tsuda, H., Maruyama, K., Nozawa, S., and Hirohashi, S. Beta-catenin mutation in carcinoma of the uterine endometrium. *Cancer Res.* 58:3526–3528, 1998.
20. Sparks, A.B., Morin, P.J., Vogelstein, B., and Kinzler, K.W. Mutational analysis of the APC/beta-catenin/Tcf pathway in colorectal cancer. *Cancer Res.* 58:1130–1134, 1998.
21. Voeller, H.J., Truica, C.I., and Gelmann, E.P. Beta-catenin mutations in human prostate cancer. *Cancer Res.* 58:2520–2523, 1998.
22. Zurawel, R.H., Chiappa, S.A., Allen, C., and Raffel, C. Sporadic medulloblastomas contain oncogenic beta-catenin mutations. *Cancer Res.* 58:896–899, 1998.
23. Molenaar, M., van de Wetering, M., Oosterwegel, M., Paterson-Maduro, J., Godsave, S., Korinek, V., Roose, J., Destree, O., and Clevers, H. Xtcf-3 transcription factor mediates b-catenin-induced axis formation in *Xenopus* embryos. *Cell* 86:391–399, 1996.
24. Behrens, J., von Kries, J.P., Kuhl, M., Bruhn, L., Wedlich, D., Grosschedl, R., and Birchmeier, W. Functional interaction of beta-catenin with the transcription factor LEF-1. *Nature* 382:638–642, 1996.
25. He, T.C., Sparks, A.B., Rago, C., Hermeking, H., Zawel, L., da Costa, L.T., Morin, P.J., Vogelstein, B., and Kinzler, K.W. Identification of c-MYC as a target of the APC pathway. [In Process Citation] *Science* 281:1509–1512, 1998.
26. Kuroda, S., Fukata, M., Nakagawa, M., Fujii, K., Nakamura, T., Ookubo, T., Izawa, I., Nagase, T., Nomura, N., Tani, H., Shoji, I., Matsuura, Y., Yonehara, S., and Kaibuchi, K. Role of IQGAP1, a target of the small GTPases Cdc42, and Rac1, in regulation of E-cadherin-mediated cell-cell adhesion. *Science* 281:832–835, 1998.
27. Hoschuetzky, H., Aberle, H., and Kemler, R. Beta-catenin mediates the interaction of the cadherin-catenin complex with epidermal growth factor receptor. *J. Cell. Biol.* 127:1375–1380, 1994.
28. Fuchs, M., Muller, T., Lerch, M.M., and Ullrich, A. Association of human protein-tyrosine phosphatase kappa with members of the armadillo family. *J. Biol. Chem.* 271:16712–16719, 1996.
29. Tao, Y.S., Edwards, R.A., Tubb, B., Wang, S., Bryan, J., and McCrea, P.D. b-catenin associates with the actin-bundling protein fascin in noncadherin complex. *J. Cell Biol.* 134:1271–1281, 1996.
30. Miyoshi, Y., Nagase, H., Ando, H., Ichii, S., Nakatsura, S., Aoki, T., Miki, Y., Mori, T., and Nakamura, Y. Somatic mutations of the APC gene in colorectal tumors: mutation cluster region in the APC gene. *Hum. Mol. Genet.* 1:229–223, 1992.
31. Peifer, M., Sweeton, D., Casey, M., and Wieschaus, E. *wingless* and zeste-white 3 kinase trigger opposing changes in the intracellular distribution of armadillo. *Development* 120:369–380, 1994.
32. Porfiri, E., Rubinfeld, B., Albert, I., Hovanes, K., Waterman, M., and Polakis, P. Induction of a beta-catenin-LEF-1 complex by wnt-1 and transforming mutants of beta-catenin. *Oncogene* 15:2833–2839, 1997.
33. Zeng, L., Fagotto, F., Zhang, T., Hsu, W., Vasicek, T.J., Perry, W.L., Lee, J.J., Tilghman, S.M., Gumbiner, B.M., and Constantini, F. The mouse Fused locus encodes Axin, an inhibitor of the Wnt signaling pathway that regulates embryonic axis formation. *Cell* 90:181–192, 1997.
34. Hart, M.J., de los Santos, R., Albert, I.N., Rubinfeld, B., and Polakis, P. Downregulation of beta-catenin by human Axin and its association with the APC tumor suppressor, beta-catenin, and GSK3 beta. [In Process Citation] *Curr. Biol.* 8:573–581, 1998.
35. Behrens, J., Jerchow, B.A., Wurtele, M., Grimm, J., Asbrand, C., Wirtz, R., Kuhl, M., Wedlich, D., and Birchmeier, W. Functional interaction of an axin homolog, conductin, with beta-catenin, APC, and GSK3beta. *Science* 280:596–599, 1998.
36. Friedl, W., Meuschel, S., Caspari, R., Lamberti, C., Krieger, S., Sengteller, M., and Propping, P. Attenuated familial adenomatous polyposis due to a mutation in the 3′ part of the APC gene. A clue for understanding the function of the APC protein. *Hum. Genet.* 97:579–584, 1996.

DIETARY LIPIDS, INFLAMMATION, AND COLON CANCER

Connye N. Kuratko

Department of Pathology
Texas Tech University Health Sciences Center
Lubbock, Texas 79430

1. INTRODUCTION

In order to improve our understanding of the process of oncogenesis in the colon and the ability to terminate or control the development and spread of tumors, we must improve our understanding of this tissue's immune system.[1] The immune system is believed to play a key role in tissue defense against tumor formation and growth.[2] Without adequate surveillance and destruction of the abnormal cell, tumors may go undetected and grow. However, hyper-immune responses and chronic inflammation may actually serve to promote tumor development in this tissue.[3] Optimal immune function must involve a balance between these two extremes. There is accumulating evidence from both human and animal studies indicating that inflammation may contribute to carcinogenesis in the colon.[3,4]

Many external factors are known to affect the efficiency of the immune system. Dietary lipids, in particular, have been shown to influence a variety of immune functions. Fatty acids from the n-6 series are necessary for T-cell mediated immune functions but they can contribute to hyper-immune responses when consumed in large amounts. Diets that contain significant levels of fatty acids from the n-3 series, in contrast, are anti-inflammatory.[2,3] Dietary lipids also influence the development of colon tumors.[5–7] Although not conclusive, diets high in n-6 fatty acids have been shown to enhance colon tumorigenesis in rats whereas diets containing n-3 fatty acids do not.[4,6] The mechanism by which n-3 fatty acids exert their anti-inflammatory, antitumor, and immunomodulating effects is unknown.[4] Iron is a second nutrient that affects immune function. It is a key factor in a number of enzymes including those involved in immune competence and cell division.[8] Some human and animal studies suggest that high levels of dietary iron or high iron status enhance the risk for colon cancer.[8–11]

Colorectal cancer remains the second leading cause of cancer death in this country, a rate that is twice that of other, less developed countries.[12] Epidemiological

Colon Cancer Prevention, edited under the auspices of AICR
Kluwer Academic / Plenum Publishers, New York, 1999.

studies have pointed to environmental factors, particularly those of the diet, that are responsible for a large portion of this increased rate.[13] The typical American diet is rich in both lipids and iron and current dietary recommendations for prevention of this disease encourage a diet that would decrease their consumption.[14] Investigations into immune variables that are sensitive to dietary modulation are important, therefore, in the study of colon cancer.

1.1. Inflammation

The majority of current literature recognizes a link between chronic inflammation and the development of cancer.[3] Inflammation is one of the necessary processes used by the body to reject unwanted tissues and invaders. Unfortunately, the inflammatory process is very non-specific and causes injury to normal host cells as well as foreign invaders. Chronic inflammation may result from infections or environmental exposures such as diet.[15] Healthy tissues chronically inundated by inflammatory cells undergo various forms of adaptation to avoid injury and death. In the colon, excessive levels of inflammation are associated with colitis, inflammatory bowel disease, and perhaps cancer.[16] Inflammation in the bowel is thought to induce cytokine release, increase free radical formation, and alter cell division by affecting both apoptotic and necrotic cell death.[3] Studies have reported an inhibition of colon carcinogenesis in response to chronic ingestion of anti-inflammatory drugs such as aspirin, piroxicam, indomethacin, and sulindac.[17]

1.2. Free Radicals

A number of studies implicate oxidative damage as a primary mechanism for inflammation-related carcinogenesis.[3] Initiation and promotion of colon tumors are closely associated with a progressive accumulation of mutations, deletions, and inappropriate repair of key genes that may increase with increased free radical production.[18] Radicals also act as regulators of cell division in the normal state, and may adversely affect the rate of cell growth, proliferation, differentiation, and apoptosis if uncontrolled.[19] At the site of inflammation in the gut, activated immune cells are capable of producing large quantities of reactive oxygen metabolites.[20] This insult to the colon results in the release of additional pro-inflammatory cytokines. These radicals and cytokines, in turn, cause induction of cytoprotective enzymes such as manganese superoxide dismutase.[20,21]

1.3. Manganese Superoxide Dismutase (MnSOD)

MnSOD is an enzyme found in the mitochondrial matrix that catalyzes the detoxification of superoxide radicals. Like most mitochondrial proteins, MnSOD is encoded by a nuclear gene, synthesized in the cytosol as a larger precursor with an N-terminal transit peptide, and subsequently transported into the mitochondria.[22] Regulation of MnSOD may be particularly critical in the colon. Contents of the colonic lumen include many foreign antigens and bacteria.[20] Disruption in the epithelial barrier of the colon allows entry of these luminal contents into the submucosa triggering an inflammatory response.[20] The colon's ability to increase MnSOD activity may be critical for surviving such inflammatory insults.

The MnSOD gene has been described as having tumor suppressive activity and is extremely low or absent in most tumor cells.[22,23] It is not clear if the decrease in MnSOD activity is a cause for, or the result of, carcinogenesis.[24] However, genetic diseases involving decreased MnSOD activity increase the risk for some cancers and increased MnSOD expression suppresses malignant characteristics of tumor cells.[22] Therefore, the literature supports the importance of MnSOD activity in various stages of this disease.

1.4. Diet and Colonic Inflammation

Diet has the potential to affect colonic inflammation through two mechanisms. First, diet may influence the production of inflammatory mediators by immune cells. Secondly, diet may influence the colon's ability to respond to those mediators. Below are reports of effects that diet has on both of these mechanisms. First, diet is shown to alter MnSOD, a cytoprotective enzyme induced by inflammation. Secondly, dietary lipids are shown to influence the structure and function of colonic immune cells.

2. EFFECTS OF DIETARY LIPIDS AND IRON ON MnSOD ACTIVITY *IN VIVO*

Both dietary fat and dietary iron have the potential for modulation of MnSOD activity and/or expression. Incorporation of polyunsaturated fatty acids into membrane phospholipids in combination with high iron levels may increase lipid peroxidation, oxygen radical formation and the level of antioxidant protection needed by the colon. This laboratory has conducted experiments to investigate the effect that dietary lipids and iron have on MnSOD activity in the colon and their implications for carcinogenesis in this tissue.

2.1. Moderate Increases in Dietary Lipid and Iron Decrease Colonic MnSOD Activity

In order to determine the effect of dietary lipid and iron on colonic mucosal MnSOD *in vivo*, two experiments were designed using weanling male Fischer rats fed experimental diets. In the first experiment, diets containing either 5% corn oil (BD), 20% corn oil (CO), 19% menhaden oil and 1% corn oil (MO), or 19% beef tallow and 1% corn oil (BT) were fed to rats for six weeks. Each of the four diets contained 35 mg/kg of iron. Four additional test diets included the same formulation plus 140 mg/kg iron. The high fat diets used in this experiment contained approximately 40% of energy from fat, or an amount that is similar to the typical Western diet. Iron was supplemented at a level approximately four times the recommended intake, a typical level of supplementation used by the public.

Table 1 shows that increasing the fat content of the diet from 20% to 40% decreased MnSOD activity in colon mucosa in rats consuming the standard AIN level of 35 mg/kg of iron. Iron supplementation further decreased MnSOD activity in animals fed the lowfat diet. Lipid peroxidation products were not increased and other antioxidant enzymes including glutathione peroxidase, catalase, and copper-zinc superoxide dismutase did not differ according to the level of lipid or iron.[25] These results indicated

Table 1. Dietary lipid and iron effects on colonic MnSOD activity[1]

Diet Goup[2]	Units/min/mg protein[3] 35 mg/kg Fe	140 mg/kg Fe
BD	3.60 ± 0.74[a]	0.90 ± 0.21[b]*
CO	2.52 ± 0.51[ab]	1.60 ± 0.45[a]
MO	1.56 ± 0.23[b]	1.61 ± 0.44[a]
BT	1.82 ± 0.30[b]	1.67 ± 0.30[a]

[1]Data taken from reference 25.
[2]BD contained 5% corn oil, CO contained 20% corn oil, MO contained 19% menhaden oil and 1% corn oil, BT contained 19% beef tallow and 1% corn oil.
[3]Means within the same column with different letters are different ($P < 0.05$). *indicates difference of iron supplemented group from nonsupplemented group.

that colonic MnSOD activity was altered by dietary patterns associated with the typical Western diet.

2.2. Colonic MnSOD Decreases during Early Weeks of Carcinogenesis

A second *in vivo* experiment was designed to further investigate dietary lipid and iron effects on colonic MnSOD and to determine the importance of this antioxidant's activity during early carcinogenesis. Weanling male Fischer rats were fed high-fat diets containing either 20% corn oil (CO), or 19% menhaden oil and 1% corn oil (MO) with a level of either 35 or 535 mg/kg iron (COFe and MOFe). Rats were given two weekly injections of either azoxymethane (AOM) at a dose of 12 mg/kg body weight or an equal volume of saline. Colonic MnSOD activity was measured 1 and 12 weeks following AOM injections.

Table 2 shows colonic MnSOD activity measured at the 1- and 12-week time points. MnSOD activity decreased significantly over the 12 week period in the carcinogen-treated animals. Iron supplementation likewise decreased MnSOD activity. The type of lipid fed did not significantly affect MnSOD.[26]

Table 2. MnSOD activity in colon mucosa following azoxymethane or saline injections[1]

	Units/min/mg protein			
	1 Week Post Injection[3]		12 Weeks Post Injection[4,5]	
Diet Group[2]	Saline	AOM	Saline	AOM
CO	9.79 ± 0.08	6.32 ± 0.76	4.40 ± 0.45	3.20 ± 1.13
MO	6.36 ± 0.62	5.93 ± 0.29	4.09 ± 1.16	3.05 ± 0.85
COFe	4.30 ± 0.66	5.14 ± 0.37	3.98 ± 0.26	2.26 ± 0.62
MOFe	4.12 ± 0.30	4.36 ± 0.09	3.00 ± 0.77	1.71 ± 0.65

[1]Data taken from reference 26.
[2]High fat modification of the AIN-93G rodent diet. CO contains 20% corn oil, MO contains 19% menhaden oil and 1% corn oil. COFe is the CO diet supplemented with iron to a level of 535 mg/kg. MOFe is the MO diet supplemented with iron to a level of 535 mg/kg.
[3]Iron supplementation decreased MnSOD activity at the one week time point ($P = 0.002$).
[4]AOM decreased MnSOD activity at the 12 week time point ($P = 0.03$).
[5]Iron supplementation decreased MnSOD activity ($P = 0.08$).

Table 3. Mineral concentration of colon mucosa[1]

Diet[2]	−Iron supplement[3]	+Iron supplement[4]
	Mucosal Manganese (μg/g wet weight)	
CO	2.25 ± 0.06	1.07 ± 0.08*
MO	1.75 ± 0.08	1.05 ± 0.06*
	Mucosal Iron (μg/g wet weight)	
CO	26.0 ± 0.69	53.8 ± 1.91*
MO	24.3 ± 0.30	61.0 ± 4.47*

[1]Data taken from reference 26.
[2]CO contains 20% corn oil, MO contains 19% menhaden oil and 1% corn oil.
[3]Diet contains 35 mg/kg iron.
[4]Diet contains 535 mg/kg iron.
*Iron supplemented group is differs from unsupplemented group ($P < 0.05$).

2.3. Iron Supplementation Decreases Concentration of Colonic Manganese

The decrease in MnSOD activity observed in rats receiving iron-supplementation may be due, in part to the effect of iron on manganese status in the colon. Table 3 shows the mineral content of colonic mucosa from rats fed the experimental diets. Mucosal iron content was significantly greater and manganese content significantly lower in animals receiving iron supplementation. Changes in MnSOD activity as the result of iron supplementation may, therefore, be primarily related to the limited availability of manganese.

A general summary of these experiments using a rat model indicates that a decrease in MnSOD activity may be an important factor in the development of colon cancer since rats treated with the colon carcinogen, AOM, showed decreased MnSOD activity during the early weeks of carcinogenesis. Dietary factors alone can modulate MnSOD activity in this tissue. Specifically, high levels of lipid and supplemental levels of iron decrease the activity of this enzyme without overtly causing damage to mucosal tissue. The mechanism involved in nutrient modulation of the enzyme is not known, but may involve limitation of manganese availability in this tissue.

3. EFFECT OF LIPIDS AND IRON ON MnSOD IN INTESTINAL CELLS *IN VIVO*

Additional experiments were designed to test the potential use of an *in vivo* model for the study of dietary effects on colonic MnSOD. Rat intestinal epithelial cells (IEC-6) were grown in media supplemented with 40 μM of either linoleic acid (LA) or eicosapentaenoic acid (EPA) and 2 mmol/L iron. MnSOD activity and protein levels were determined in the treated cells following 8 hour exposure to 10 ng/ml tumor necrosis factor-α (TNF). TNF is a primary mediator of inflammation in the colon and a known inducer of MnSOD in these cells.[27]

3.1. Effect of n-6 and n-3 Fatty Acids on TNF Induction of MnSOD

Figure 1 shows MnSOD activity in IEC-6 cells as the result of fatty acid supplementation.[28] MnSOD activity increased in these cells as the result of fatty acid supple-

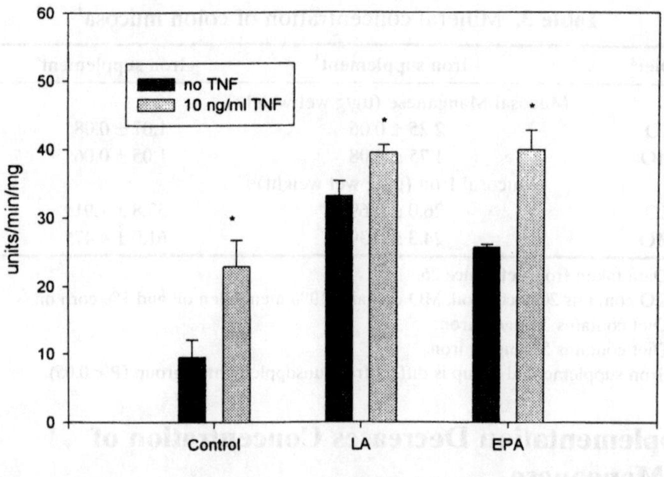

Figure 1. MnSOD activity in IEC-6 cells grown in fatty acid-supplemented media and treated ±10 ng/ml tumor necrosis factor-α (TNF) for 8 hours. Control media is DMEM supplemented with 0.6 mg/ml L-glutamine, 10 μg/ml insulin, and 10% fetal bovine serum; LA and EPA treatments were control media supplemented with 40 μmol/L linoleic acid or eicosapentaenoic acid. Values represent Mean ± SEM of six culture plates. Without TNF, LA > EPA > control ($P < 0.05$). With TNF, LA and EPA > control ($P < 0.05$) *indicates difference from non-TNF-treated group. Data taken from reference 28.

mentation, particularly linoleic acid. The increased activity of MnSOD seen in the LA-supplemented group may indicate a direct influence of the fatty acid on MnSOD or it may be an indicator of an increase in production of inflammatory cytokines. In contrast, the lower MnSOD activity in EPA-treated cells may reflect an anti-inflammatory environment. The ability of LA to influence inflammatory mediators, specifically in colon cells, may serve as a mechanism of tumor promotion.

Treatment of IEC-6 cells with TNF increased MnSOD activity in all groups as shown in Fig. 1. This is in agreement with previous work in IEC-6 cells.[27] The type of lipid used did not alter the ability of cells to increase MnSOD activity in response to TNF.

Figure 2 shows the level of MnSOD protein produced in IEC-6 cells treated with fatty acids and TNF. TNF treatment increased the level of MnSOD produced in all groups. Although lipid supplementation increased MnSOD activity, detectable protein production did not change from control.

3.2. Effect of Iron Supplementation on TNF Induction of MnSOD

A similar experiment was designed to investigate the effect of iron on MnSOD activity in IEC-6 cells.[29] Ferrous sulfate was added to cell culture media at a level that provided 0.5, 1.0, or 2.0 mmol/L iron. At the 2.0 mmol/L level, iron increased lipid peroxidation products, but did not increase the rate of proliferation in the IEC-6 cells (data not shown). Figure 3 shows that iron supplementation increased MnSOD activity and protein production in all groups. Figure 4 shows MnSOD activity was increased by TNF, but that iron did not further increase the activity in TNF-treated cells.

It is important to note several differences in MnSOD response to lipids and iron using the cell culture model as compared to the *in vivo* model. First the overall MnSOD

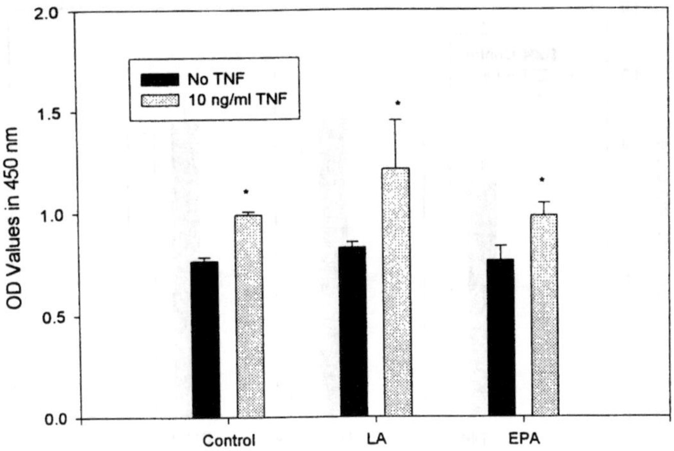

Figure 2. MnSOD protein levels in IEC-6 cells determined by ELISA. Control media is DMEM supplemented with 0.6 mg/ml L-glutamine, 10 μg/ml insulin, and 10% fetal bovine serum; LA and EPA treatments were control media supplemented with 40 μmol/L linoleic acid or eicosapentaenoic acid. There were no differences according to fatty acid supplementation ($P > 0.05$). *indicates significant increase as the result of treatment with TNF ($P < 0.05$).

activity in rat colonic mucosa was lower than in the IEC-6 cultures under all conditions. Second, both lipid and iron supplementation decreased MnSOD activity in the *in vivo* model whereas lipids and iron increased MnSOD activity in IEC-6 cells. Previous studies have proposed that exposure to environmental metals may effect MnSOD by a combination of factors.[30] First, there may be an induction of MnSOD gene expression as the result of metal-induced oxidative stress. Secondly there appear to be post-translational modifications of MnSOD that occur in response to metal exposure. In the current study, iron supplementation increased lipid peroxidation in IEC-6 cells whereas

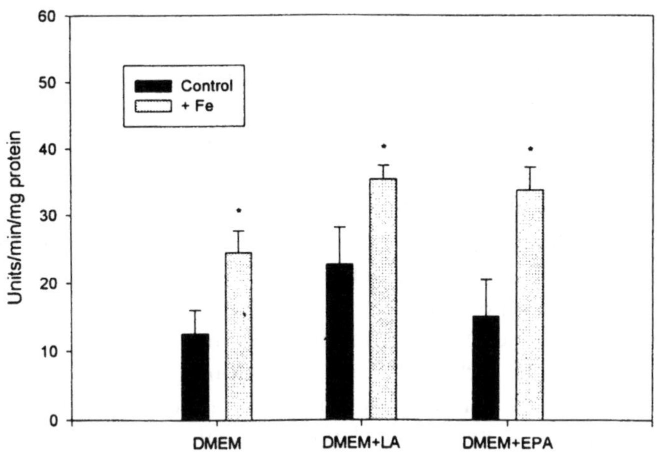

Figure 3. MnSOD activity in IEC-6 cells grown in fatty acid-supplemented media ±2 mmol/Fe. Control media is DMEM supplemented with 0.6 mg/ml L-glutamine, 10 μg/ml insulin, and 10% fetal bovine serum; LA and EPA treatments were control media supplemented with 40 μmol/L linoleic acid or eicosapentaenoic acid. Values represent Mean ± SEM of six culture plates. *indicates significant increase as the result of iron supplementation ($P < 0.05$). Data taken from reference 29.

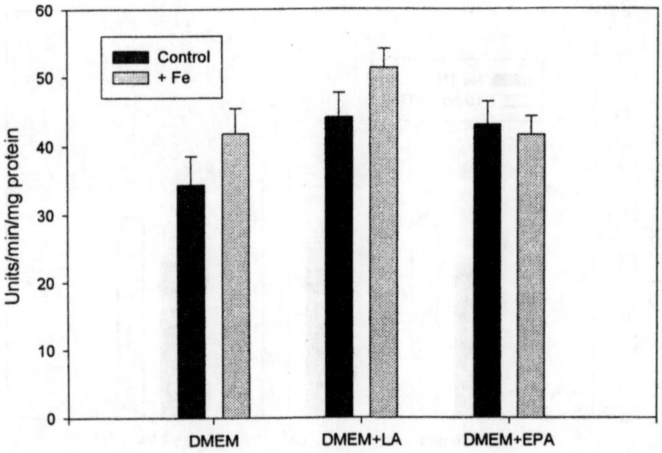

Figure 4. MnSOD activity in IEC-6 cells grown in fatty acid-supplemented media ±2mmol/Fe and treated with 10ng/ml tumor necrosis factor-α (TNF) for 8 hours. Control media is DMEM supplemented with 0.6 mg/ml L-glutamine, 10μg/ml insulin, and 10% fetal bovine serum; LA and EPA treatments were control media supplemented with 40μmol/L linoleic acid or eicosapentaenoic acid. Values represent Mean ± SEM of six culture plates. *indicates significant increase as the result of treatment with TNF ($P < 0.05$). Data taken from reference 29.

colon mucosa has been shown to be resistant to peroxidation.[31] Therefore, it is possible that differences in oxidative stress imposed by iron resulted in variations of MnSOD in these two models. Differences in MnSOD regulation may also occur as the result of limited manganese availability and post-translational MnSOD changes in the *in vivo* model. Future studies of the effect of iron on colonic MnSOD using both systems should address whether changes in activity occur as the result of increased mRNA levels or as the result of post-translational alterations due to protein stability or limited manganese availability.

4. EFFECT OF DIETARY LIPIDS ON COLONIC LYMPHOCYTES

Peripheral blood has been used almost exclusively as the source of cells for most immunological studies including those investigating gastrointestinal cancer.[32] The circulating lymphoid cells are used because of their availability and relative ease of preparation.[33] However, immune responses in the colon may be influenced more directly by local lymphocytes than by those that circulate. The mucosal immune system, including the colon, is a thymus-independent site of T-cell development. Lymphocytes in the intestinal mucosa have a unique pattern of maturation and differ from circulating lymphocytes in many respects including T-cell functions and cell proliferation.[34–36] As primary mediators in mucosal immunity, colonic lymphocytes (CL) may play an important role in both the protective and pathological consequences of colon tissue inflammation.

Mucosal lymphocytes, including those of the colon are capable of producing cytokines and other factors that influence inflammation in the tissue. They maintain a state of down regulation under normal conditions to prevent hyper-immune response to the tremendous number of antigens that are routinely exposed to the colon.[16] Dietary

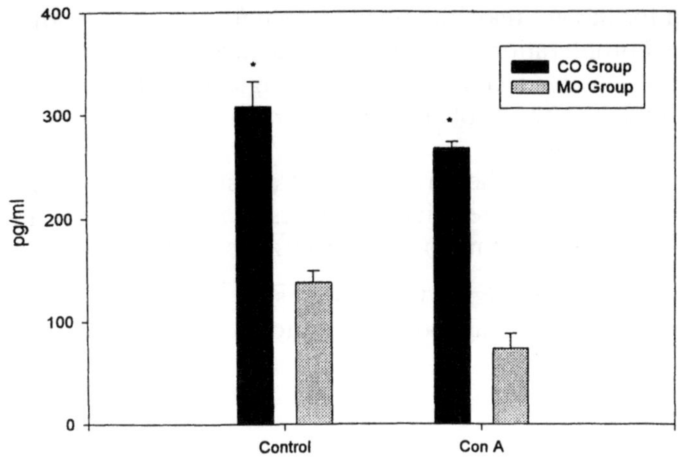

Figure 5. Prostaglandin E_2 production by colonic lymphocytes from rats fed diet high in either corn oil or menhaden oil. PGE_2 was measured in media from colonic lymphocytes after 72 h in culture. Values are means ± SEM of 3 rats. Control received no mitogen stimulation; Con A received 6.25 μg concavalin A/ml. *indicates CO > MO ($P < 0.05$). Data taken from reference 37.

lipids are known to affect functions of peripheral lymphocytes, but effects on CL are unknown. Dietary effects on the activity of these unique and tissue specific cells should be investigated in studies of inflammation and cancer in the colon.

This laboratory conducted a study designed to determine the effect of n-3 and n-6 fatty acids on CL. Male Fischer rats were fed either the CO or MO diet mentioned previously. CL were isolated by a modification of the procedure by Bull and Bookman[33] and cultured in the presence or absence of mitogen stimulation for 72 hours.

Results showed that dietary lipids affect both the structure and function of CL.[37] Fatty acid composition of CL reflected that of the diet. LA content of CL from corn oil fed rats was 21.9% compared to 11.3% in menhaden oil fed rats. Arachidonic acid levels were 5.4% in the corn oil group compared to 2.4% in the menhaden oil group. Figure 5 shows that prostaglandin E_2 (PGE_2) production was higher in CL of corn oil fed rats than of menhaden oil fed rats. At high levels, PGE_2 has pro-inflammatory functions in many tissues including the colon.[17] Lipids, therefore, are able to modulate the production of inflammatory mediators by immune cells in the colon.

5. SUMMARY AND FUTURE DIRECTIONS

Chronic inflammation has been associated with diseases of the bowel including cancer. High levels of dietary lipid and iron are associated with tumor development in the colon. Lipid and iron involvement in chronic inflammation may be a mechanism whereby diet influences colon tumorigenesis. Studies from this laboratory show involvement of dietary factors in: 1) the production of inflammatory mediators in the colon, and 2) the response of colonocytes to inflammation. Specifically, results have shown:

- Dietary lipid and iron, at levels typically consumed in the typical Western diet decrease the activity of colonic MnSOD. This decrease in antioxidant pro-

tection for the gut may increase the potential damage from oxidative stress, alter cell proliferation or apoptotic pathways.
- MnSOD activity decreases in early stages of colon cancer. Decreases in MnSOD as the result of dietary factors may promote the development of cancer in this tissue.
- Dietary lipids alter the structure and prostaglandin production of colonic lymphocytes. Therefore, dietary factors may play a part in the type and potency of pro-inflammatory products secreted by colonic immune cells.

Future studies will investigate the relationship of colonic lymphocytes to other mucosal cells during both normal and inflammatory conditions.

ACKNOWLEDGMENTS

The author thanks the American Institute for Cancer Research and the South Plains Foundation for grants in support of this research.

REFERENCES

1. Lowell, J., Parnes, H., and Blackburn, G. Dietary immunomodulation: Beneficial effects on oncogenesis and tumor growth. *Critical Care Med.* 18(2):S145–148, 1990.
2. Maki, P. and Newberne, P. Dietary lipids and immune function. *J. Nutr.* 122:610–614, 1992.
3. Meeting Report. Molecular mechanisms for inflammation-promoted pathogenesis of cancer- the sixteenth international symposium of the Sapporo cancer seminar. *Cancer Res.* 57:3620–3624, 1997.
4. Singh, J., Hamid, R., and Reddy, B. Dietary fat and colon cancer: Modulation of cyclooxygenase-2 by types and amount of dietary fat during the postinitiation stage of colon carcinogenesis. *Cancer Res.* 57:3465–3470, 1997.
5. Doll, R. and Peto, R. The causes of cancer: quantitative estimates of avoidable risks of cancer. *J. Natl. Cancer Inst.* 66:1191–1308, 1981.
6. Reddy, B. and Sugie, S. Effect of different levels of omega-3 and omega-6 fatty acids on azoxymethane-induced colon carcinogenesis in F344 rats. *Cancer Res.* 48:6642–6647, 1988.
7. Willett, W., Stampfer, M., Colditz, G., Rosner, B., and Speizer, F. Relation of meat, fat, and fiber intake to the risk of colon cancer in a prospective study among women. *New Engl. J. Med.* 323:1664–1672, 1990.
8. Nelson, R. Dietary iron and colorectal cancer risk. *Free Rad. Biol. Med.* 12:161–168, 1992.
9. Nelson, R., Davis, F., Sutter, W., Sobin, L., Kikendall, J., and Bowen, P. Body iron stores and risk of colonic neoplasia. *J. Natl. Cancer Inst.* 86:455–460, 1994.
10. Wurzelmann, J., Silver, A., Schreinemachers, D., Sandler, R., and Everson, R. Iron intake and the risk of colorectal cancer. *Cancer Epidemiol. Biomarkers Prev.* 5:503–507, 1996.
11. Bird, C., Witte, J., Swendseid, M., Shikany, J., Hunt, I., Frankl, H., Lee, R., Longnecker, M., and Haile, R. Plasma ferritin, iron intake, and the risk of colorectal polyps, *Am. J. Epidemiol.* 144:34–41, 1996.
12. Miller, B.A., Ries, L.A., Hankey B.F., Kosary, C.L., and Edwards, B.K. (eds). Cancer Statistics Review: 1973–1989, p. VI.1, National Cancer Institute, NIH Pub. No. 92-2789, 1992.
13. Wargovich, M. and Mastromarino, A. Dietary factors in the etiology and prevention of colon cancer. *The Ca Bulletin* 46:303–308, 1994.
14. World Cancer Research Fund and American Institute for Cancer Research. *Food, Nutrition and the Prevention of Cancer: a Global Perspective*, American Institute for Cancer Research, 1997.
15. Kinsella, J. and Lokesh, B. Dietary lipids, eicosanoids, and the immune system. *Crit. Care Med.* 18(2):S94–113, 1990.
16. Neurath, M. and Buschenfelde, M. Protective and pathogenesis roles of cytokines in inflammatory bowel diseases. *J. Invest. Med.* 44:516–521, 1996.
17. DeBois, R., Giardiello, F., and Smalley, W. Nonsteroidal anti-inflammatory drugs, eicosanoids, and colorectal cancer prevention. *Gastroenterol. Clinics N. America* 25:773–791, 1996.

18. Augenlicht, L., Velcich, A., and Heerdt, B. Short-chain fatty acids and molecular and cellular mechanisms of colonic cell differentiation and transformation. *Adv. Exp. Med. Biol.* 375:137–148, 1995.
19. McCord, J. Superoxide radical: controversies, contradictions, and paradoxes. *Proc. Soc. Exp. Biol. Med.* 209:112–117, 1995.
20. Tannahill, C., Stevenot, S., Eaker, E., Sallustio, J., Nick, H., and Valentine, J. Regulation of superoxide dismutase in primary cultures of rat colonic smooth muscle cells. *Am. J. Physiol.* 272:G1230–G1235, 1997.
21. Ogle, C., Mao, J., We, J., Ogel, J., and Alexander, J. The 1994 Lindberg award: The production of tumor necrosis factor, interleukin-1, interleukin-6, and ptrostaglandin E_2 by isolated enterocytes and gut macrophages: Effect of lipopolysaccharide and thermal injury. *J. Burn Care Rehab.* 15:470–477, 1994.
22. Church, S., Grant, J., Ridnour, L., Oberley, L., Swanson, P., Meltzer, P., and Trent, J. Increased manganese superoxide dismutase expression suppresses the malignant phenotype of human melanoma cells. *Proc. Natl. Acad. Sci. USA* 90:3113–3117, 1993.
23. St. Clair, C. and Holland, J. Complementary DNA encoding human colon cancer manganese superoxide dismutase and the expression of its gene in human cells. *Cancer Res.* 51:939–943, 1991.
24. Nelson, R. Superoxide dismutase in cultured benign and malignant tumors of the colon. *Basic Life Sci.* 49:699–702, 1988.
25. Kuratko, C. Increasing dietary lipid and iron content decreases manganese superoxide dismutase activity in colonic mucosa. *Nutr. Cancer* 28:36–40, 1997.
26. Kuratko, C. Decrease of manganese superoxide dismutase activity in rats fed high levels of iron during colon carcinogenesis. *Food Chem. Toxicol.* 36:819–824, 1998.
27. Valentine, J. and Nick, H. Acute-phase induction of manganese superoxide dismutase in intestinal epithelial cell lines. *Gastroenterol.* 103:905–912, 1992.
28. Kuratko, C. and Constante, B. Linoleic acid and tumor necrosis factor-alpha increase manganese superoxide dismutase activity in intestinal cells. *Cancer Lett.* 130:191–196, 1998.
29. Kuratko, C.N. Iron increases manganese superoxide dismutase activity in tumor necrosis factor-α stimulated IEC-6 cells. *Toxicol. Lett.* (in press), 1998.
30. Kumagi, Y., Mizukado, S., Nagafune, J., Shinyashiki, M., Homma-Takeda, S., and Shimojo, N. Post-transcriptional elevation of mouse brain MnSOD protein by mercuric chloride. *Brain Res.* 769:178–182, 1997.
31. Kuratko, C.N., Tsai, S.Y., and Pence, B.C. Effects of dietary fat and 1,2-dimethylhydrazine on microsomal lipid peroxidation. *J. Nutr. Biochem.* 5:78–83, 1994.
32. Endre, S., Eisenhut, T., and Sinha, B. n-3 Polyunsaturated fatty acids in the regulation of human cytokine synthesis. *Biochem. Soc. Trans.* 23:277–281, 1994.
33. Bull, D. and Bookman, M. Isolation and functional characterization of human intestinal mucosal lymphoid cells. *J. Clin. Invest.* 59:966–974, 1977.
34. Elitsur, Y., Bull, A., and Luk, G. Modulation of human colonic lamina propria lymphocyte proliferation: effect of bile acids and oxidized fatty acids. *Dig. Dis. Sci.* 35:212–220, 1990.
35. Poussier, P., Edouard, P., Leed, C., Binnie, M., and Julius, M. Thymus-independent development and negative selection of T cells expressing T cell receptor a/b in the intestinal epithelium: evidence for distinct circulation patterns of gut and thymus derived T lymphocytes. *J. Exp. Med.* 176:187–199, 1992.
36. Doe, W. The intestinal immune system. *Gut* 30:1679–1685, 1989.
37. Kuratko, C. and Becker, S. Dietary lipids alter fatty acid composition and PGE_2 production in colonic lymphocytes. *Nutr. Cancer* 31:56–61, 1998.

5

SULINDAC SULFONE INDUCED REGRESSION OF RECTAL POLYPS IN PATIENTS WITH FAMILIAL ADENOMATOUS POLYPOSIS

G. D. Stoner,[1,2,3]* G. T. Budd,[4] R. Ganapathi,[4] B. DeYoung,[3]
L. A. Kresty,[1] M. Nitert,[1] B. Fryer,[5] J. M. Church,[4] K. Provencher,[4]
R. Pamukcu,[5] G. Piazza,[5] E. Hawk,[6] G. Kelloff,[6] P. Elson,[5] and
R. U. van Stolk[4]

[1]Ohio State University School of Public Health
[2]Comprehensive Cancer Center and [3]Department of Pathology
Columbus, Ohio
[4]Cleveland Clinic Foundation
Cleveland, Ohio
[5]Cell Pathways, Inc.
Horsham, Pennsylvania
[6]National Cancer Institute
Bethesda, Maryland

1. ABSTRACT

Sulindac sulfone (Exisulind), a metabolite of the non-steroidal anti-inflammatory drug, sulindac, was evaluated for its effects on the development of rectal polyps in patients with familial adenomatous polyposis. Three cohorts of 6 patients each were given doses of 200, 300, or 400 mg Exisulind twice daily. Hepatotoxicity, shown by elevation in blood transaminase levels, was the dose-limiting toxicity and occurred at the 400 mg bid dose. Due to this toxicity, all patients treated with the 400 mg dose were subsequently reduced to the 200 mg dose level. Subsequently, 2 of the 6 patients were dose-escalated to 400 mg bid dose. The patients were treated with Exisulind for a period of

* Address reprint requests to: Gary D. Stoner, Ph.D., Division of Environmental Health Sciences, School of Public Health, CHRI, Rm. 1148, 300 W. 10th Avenue, Columbus, Ohio 43210

Colon Cancer Prevention, edited under the auspices of AICR
Kluwer Academic / Plenum Publishers, New York, 1999.

six months. Sixteen of 18 patients had regression of small polyps (≥6 mm in diameter) characterized by a flattening of the polyps and a macular "halo" appearance. Histopathologic examination of the polyp biopsy specimens showed a marked increase in the proportion of mucin producing cells in the glands after treatment with Exisulind at all dose levels. Ki-67 staining, a measure of cell proliferation, was higher in the polyps than in normal mucosa. There was no significant change in the proliferation index between baseline and six month values in any of the groups treated with Exisulind or in normal tissues. The median apoptotic labeling index, as determined by the TUNEL technique, was higher in the polyps than in normal-appearing mucosa. Overall, there was no significant change in the apoptotic labeling index between baseline and 6 months in normal-appearing mucosa however, the index in polyps was increased. These results suggest that treatment of FAP patients with Exisulind for a period of six months may lead to regression of small polyps, and that the mechanisms of Exisulind—induced regression appear to be through stimulation of mucus differentiation and apoptosis in glandular epithelium.

2. INTRODUCTION

Familial adenomatous polyposis (FAP) is an inherited disorder characterized by the development of hundreds to thousands of colorectal adenomatous polyps in childhood and early adulthood with inevitable progression to cancer.[1] The disease is inherited in an autosomal dominant pattern with greater than 90 percent penetrance. FAP arises from mutations in the APC tumor suppressor gene located on chromosome 5q, and all patients with FAP have a mutation in one allele of the gene. Genetic markers are now sufficiently specific that virtually all individuals in affected families can be diagnosed by genetic testing.[2] Standard management of patients with FAP includes total colectomy or subtotal colectomy in the second or third decade of life. Although rectal polyps often regress following subtotal colectomy, new polyps almost always emerge and a significant cancer risk exists. Periodic sigmoidoscopy every six to twelve months is required for the lifetime of such patients.

Nonsteroidal anti-inflammatory drugs (NSAIDs) were initially suggested as a possible therapy for colonic polyposis in FAP patients because they inhibit prostaglandin production, and some colonic tumors had been observed to produce excessive quantities of prostaglandins.[3] In 1989, Waddell et al.[4] published the results of a trial in which indomethacin and then sulindac were administered to patients with FAP. Indomethacin appeared to have little effect while sulindac caused the near total regression of colonic polyps in ten FAP patients. Follow-up of these patients revealed no further development of colon cancer or polyps unless the sulindac was discontinued. Similar findings with sulindac in FAP patients have been published by others.[5–7] Animal studies have similarly demonstrated that sulindac inhibits the development of chemically-induced small bowel and colon tumors. Two studies with sulindac in mice and rats treated with the carcinogen dimethylhydrazine demonstrated at least a 50% reduction in the development of colonic tumors and adenomas when compared to carcinogen controls.[8,9]

Although sulindac treatment results in a marked regression of polyps in FAP patients, it is not without side effects. Sulindac is metabolized into sulindac sulfide which has been shown to inhibit cyclooxygenases 1 and 2 (Cox-1 and Cox-2).[10] Inhibition of cyclooxygenases results in lowered levels of prostaglandins in gastric mucosa which leads to irritation and ulceration. The severity of these lesions precludes the use of sulindac in the long-term management of FAP. The other metabolite of sulindac is sulindac sulfone (Exisulind), which is being developed for the treatment of FAP. Exisulind lacks the antiprostaglandin synthetase activity typically associated with the gastrointestinal toxicity produced by the NSAIDs.[11] In addition, Exisulind, like sulindac, significantly inhibits the growth of colonic tumor cells *in vitro* and chemically-induced colon cancer in animals.[12,13] Based upon these observations, we decided to evaluate the effects of Exisulind on the development of colonic adenomas in FAP patients. This was undertaken as part of a Phase I clinical trial to assess the toxicity and pharmacokinetics of Exisulind in humans.

3. METHODS

This study was a single-center, dose-escalation outpatient study of 6 months duration with an extension arm. Data described here are from the first 6 months of the study. Six patients in each group were administered one of three oral doses (200, 300, and 400 mg) of Exisulind twice daily. A conventional Phase I trial design was used. Criteria for inclusion of patients in the trial were: subtotal colectomy more than three years prior to trial entry; presence of ≥5 rectal polyps on sigmoidoscopy; ≥18 years of age; no NSAIDs for at least two weeks prior to study entry; non-pregnant and practicing contraception; and informed consent. Exclusion criteria included: known hypersensitivity to NSAIDs; refractory to prior sulindac treatment; active peptic ulcer disease; known malabsorption syndromes; planned proctectomy prior to study completion; use of sulindac sulfone within one month; evidence of hepatic, renal or hematologic dysfunction; and prior malignancy.

Patients were evaluated for toxicity by telephone contact daily for the first week, weekly for the next five weeks, and monthly thereafter; physical examination at pretreatment and at 1, 4, and 6 months (off study); clinical chemistry and hematological profiles at pretreatment, at weeks 1, 2, and 4, and at months 2–6; and an upper endoscopy at pretreatment and at month 6. Asymptomatic transaminase elevation was the dose-limiting toxicity and occurred at the 400 mg bid dose. Three of six patients treated at the 400 mg dose developed grade 2–3 hepatic toxicity, according to the NCI common toxicity criteria and a fourth developed grade 1 hepatotoxicity. Hepatotoxicity resolved with stopping or lowering the dose. The safety and tolerability of Exisulind in this trial has been described.[14] Patients underwent flexible sigmoidoscopy of the rectal segment to monitor polyp size and to determine polyp number at baseline and at 1, 4, and 6 months. Normal-appearing mucosa and polyps biopsied at the time of sigmoidoscopy were examined histologically for cellular and glandular morphology, and by histochemistry to determine the proliferative and apoptotic rates in crypt cells. The proliferative index was measured by immunohistochemical staining of tissues for Ki-67 expression using a mouse monoclonal antibody to MIB-1 (Immunotech, Marseille, France) as previously described.[15] The percent Ki-67 positive nuclear area over total nuclear area was determined in normal-appearing tissue both distant to and adjacent to the polyp, and in polyp tissues from each patient using computer-assisted image analysis (Roche; Elon College, NC). Adjacent sections from the tissue blocks were evaluated for apoptosis using a modification of the TUNEL technique[16] and reported as the apoptotic index.

4. RESULTS

4.1. Polyp Counts

Polyp numbers increased during the six months treatment period in patients receiving Exisulind at a dose of 200 mg bid. In contrast, in patients given the 300 mg and 400/200 bid dose levels, the number of polyps seen during the six month period remained stable (Fig. 1). Sixteen of the eighteen patients had regression of small polyps ≤6 mm in size. Regression was also characterized by loss of volume and a "flattening" of the polyps. There was also a macular "halo" appearance of some polyps noted prominently in subjects treated with Exisulind. Biopsies of these haloed polyps indicated that they still had some identifiable adenomatous glands which showed little evidence of mucus differentiation, and multiple glands in which mucus differentiation was a prominent feature (Figs. 2 and 3). Thus, it appeared that treatment with Exisulind stimulated mucus differentiation in cells of adenomatous glands.

4.2. Cell Proliferation

Expression of the proliferation marker, Ki67, was higher in adenomatous tissue than in adjacent normal tissue (Figs. 4 and 5). Treatment with Exisulind for a period of at least six months did not lead to a significant change in Ki67 expression in either normal-appearing mucosa or in adenomatous polyps (Fig. 6).

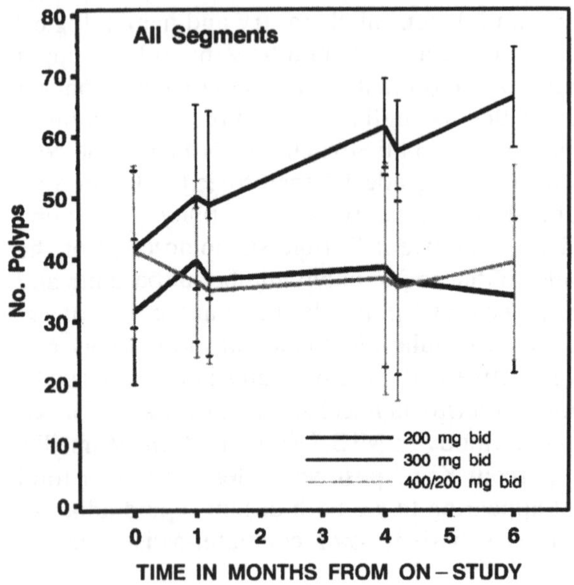

Figure 1. Effect of FGN-1 on polyp development in all segments of the rectum in FAP patients over a period of six months. Note that the polyp number increases steadily in patients treated with 200 mg bid and is stabilized in patients administered either 300 mg or 400/200 mg bid. (Reprinted courtesy of *Clinical Cancer Research*.)

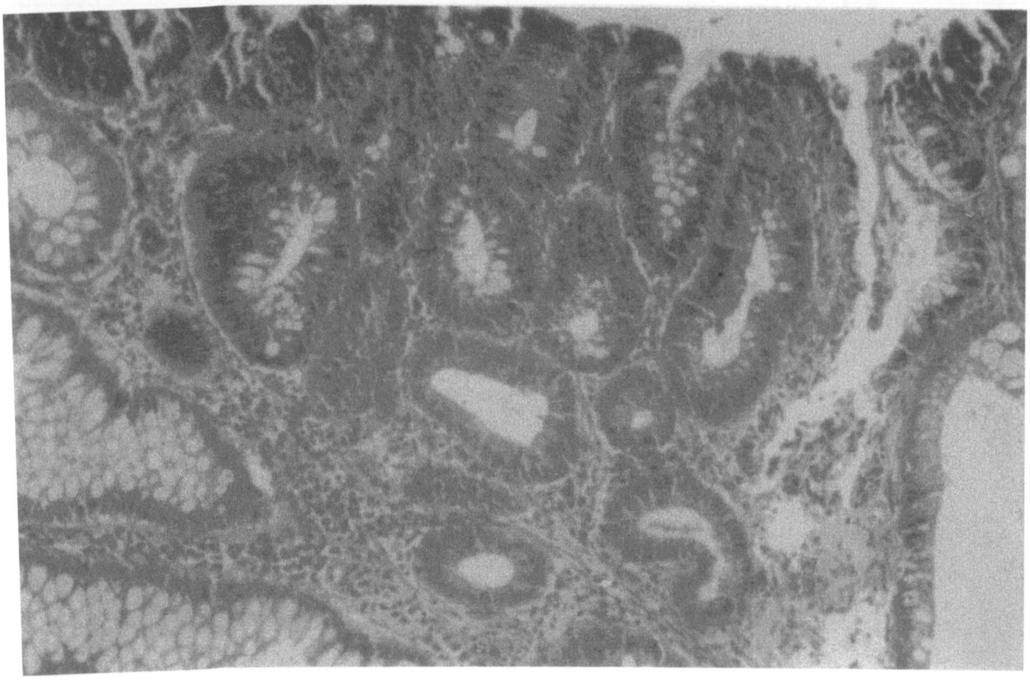

Figure 2. Adenomatous polyp taken from a patient before treatment with FGN-1. Note that most glands exhibit minimal evidence of mucus differentiation.

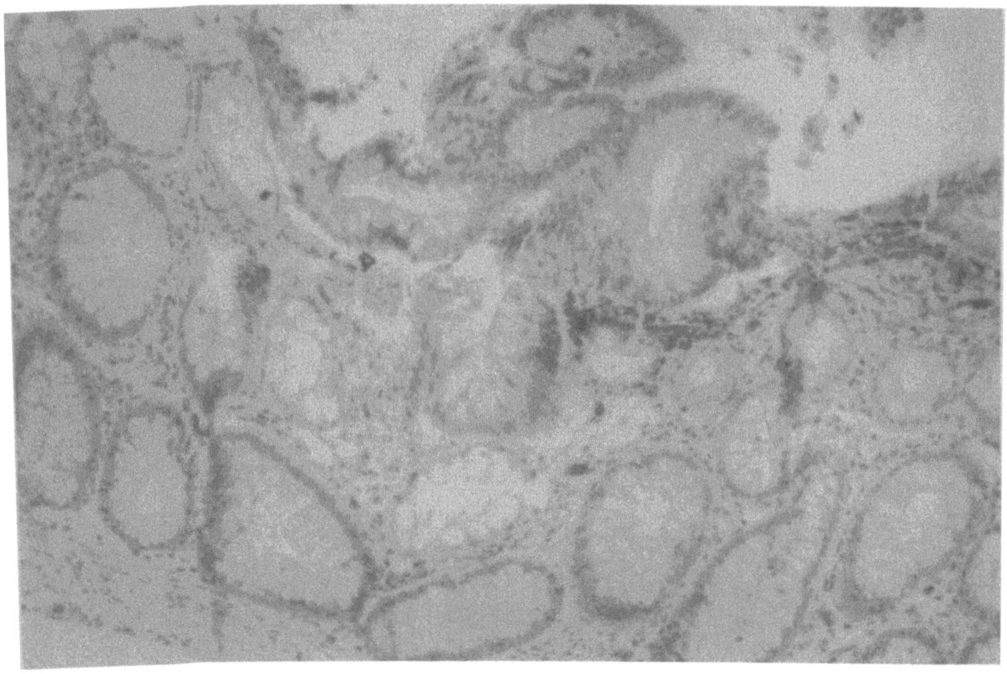

Figure 3. Adenomatous polyp taken from a patient after six months of treatment with 300 mg FGN-1 bid. Note that the dysplastic glands in the upper portion of the polyp have extensive evidence of mucus differentiation.

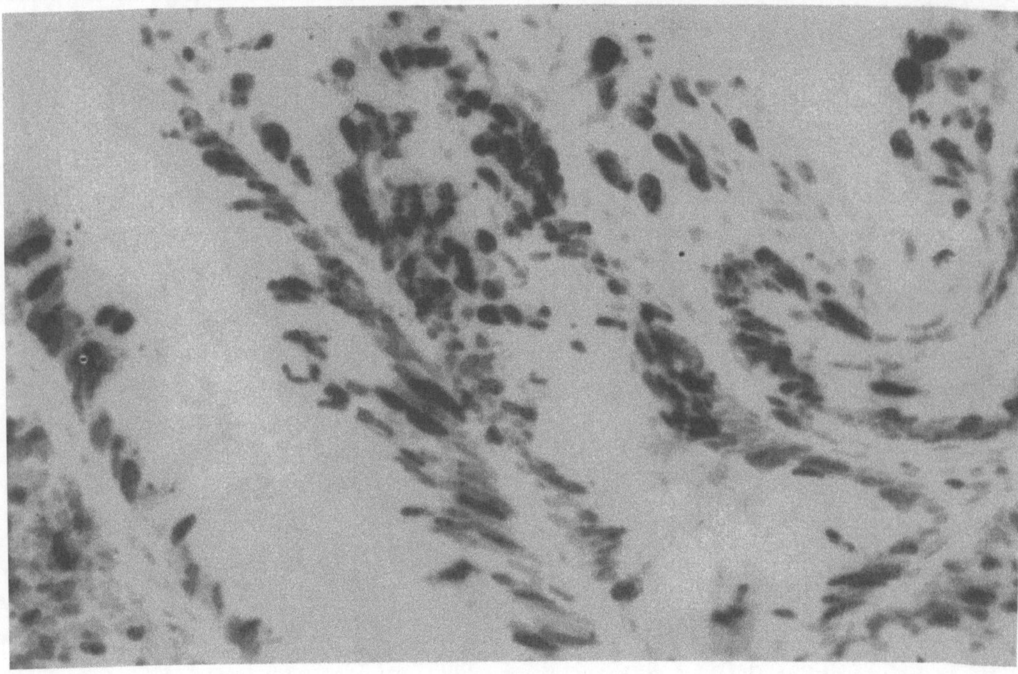

Figure 4. Histochemical staining for Ki67 in polyp tissue. Note the large number of positively stained nuclei.

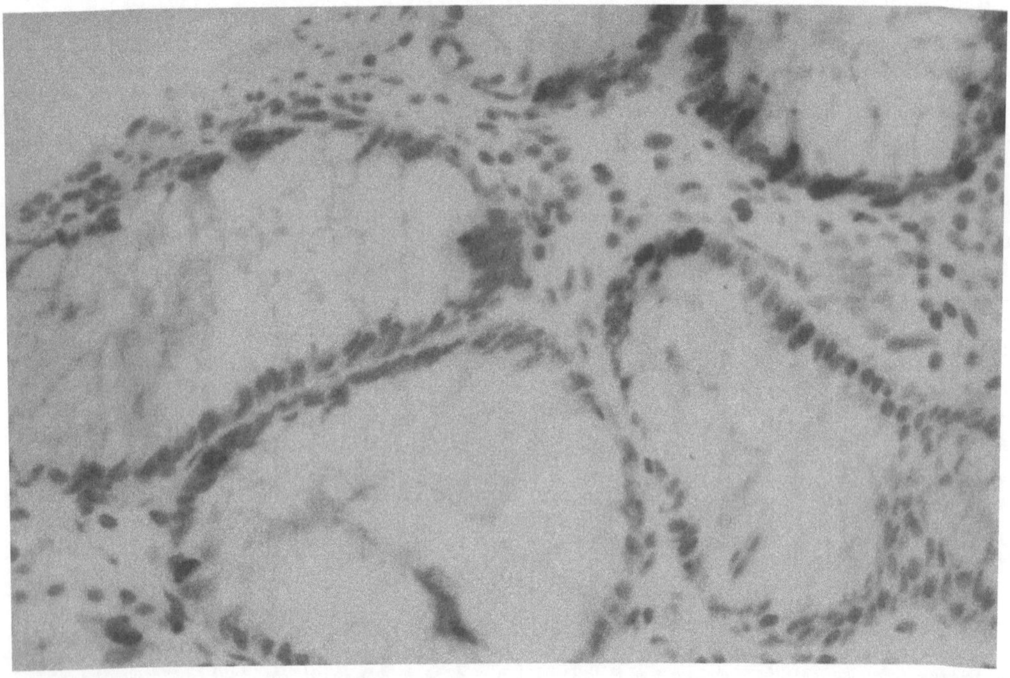

Figure 5. Histochemical staining for Ki67 in normal colonic mucosa. Note the smaller number of positively stained nuclei when compared to polyp tissue.

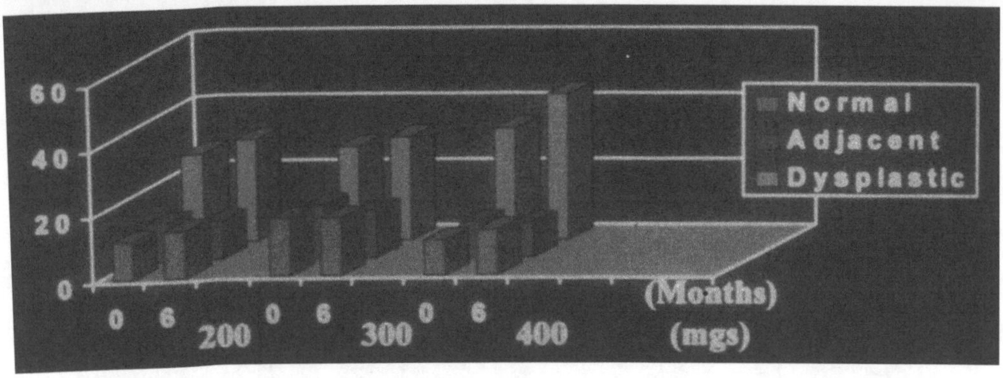

Figure 6. Percent Ki67 positive staining in normal, adjacent normal, and polyp tissues of FAP patients treated with FGN-1. Note the higher level of staining in dysplastic polyps than in adjacent normal and normal mucosa. In addition, note that treatment with FGN-1 at 300 mg bid does not influence the proliferation index in any of the tissues during a period of six months. The proliferation index in dysplastic tissues in patients treated for six months with 400 mg FGN-1 bid is not significantly higher than in controls at time zero.

4.3. Apoptosis

The median apoptotic labeling index for each patient was calculated for both normal mucosa and polyps at baseline and after six months of treatment with Exisulind. At both timepoints, the apoptotic labeling index was higher in polyps than in normal mucosa (data not shown). There was no significant change in the apoptotic labeling index of normal mucosa between baseline and month six. There was an increase however, in overall apoptotic labeling index in polyps at six months relative to baseline (Fig. 7). These data suggest that Exisulind increases the rate of apoptosis in polyps, which may serve as a mechanism for their regression.

5. DISCUSSION

This report describes the effect of sulindac sulfone (Exisulind) on the development of rectal polyps in patients with familial adenomatous polyposis. Our study revealed that treatment of FAP patients for six months with Exisulind at 300 or 400 mg bid leads to a stabilization of polyp number and consistent evidence of regression of smaller polyps. The regressed polyps were flattened and "haloed" and showed evidence of both mucus differentiation and increased apoptosis. Although not proven conclusively, the "haloed" appearance of these polyps may have been due to their extent of mucus differentiation. There was no evidence of drug-related effects on polyp number or appearance in patients treated with Exisulind at the dose of 200 mg bid. The parent compound, sulindac, at 150 mg bid, appeared to cause almost total regression of all rectal polyps within six months, suggesting that it is a more potent inhibitor of polyp development than Exisulind.[4–7] The increased potency of sulindac may be due its inhibitory effects on cyclooxygenase activity (which would result in lowered levels of growth-promoting prostaglandins) as well as its ability to stimulate cellular apoptosis.

An advantage of Exisulind compared to sulindac however, is the observation that Exisulind does not inhibit prostaglandin synthesis[11,13] which can lead to gastrointestinal

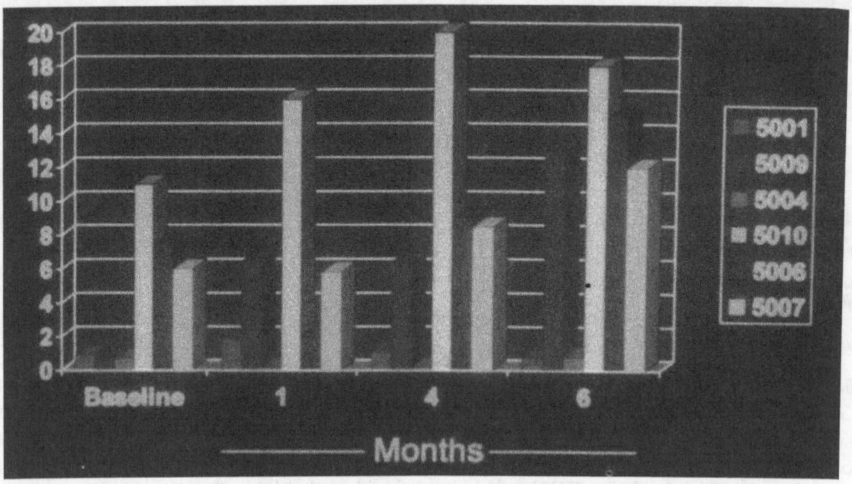

Figure 7. Percent apoptotic cells in polyps from six patients before and during treatment with FGN-1 at 300 mg bid. Each bar represents the mean percent apoptotic cells in 12–18 polyps harvested from each of the six patients at baseline and at 1, 4, and 6 months. Note that treatment with FGN-1 increased the percent apoptotic cells in polyps taken from four of the six patients (5009, 5010, 5006, 5007). The apoptotic index was not changed in polyps taken from patients 5001 and 5004.

disturbances. Gastrointestinal ulceration was not observed in patients treated during six months with Exisulind although, hepatotoxic effects were seen in two patients treated with 400 mg bid. These hepatotoxic effects resolved upon lowering the dose to 200 mg bid.

The observation of a higher level of expression of Ki67 in adenomatous tissue than in normal mucosa at baseline and after treatment with Exisulind was not unexpected since cellular proliferation in dysplastic tissues is usually higher than in normal tissues. However, the finding that Exsulind had no effect on cellular proliferation in adenomatous tissues during treatment was unexpected, particularly in view of its ability to stimulate mucus differentiation. Usually, the process of cellular differentiation is associated with a reduced rate of proliferation. However, the lack of effect on proliferation is consistent with previous cell culture studies showing that Exisulind did not inhibit cell proliferation under basal conditions.[12,17]

At the dose of 300 mg bid, treatment with Exisulind was associated with stimulation of apoptosis in adenomatous tissue. This was noted particularly in polyps that were undergoing regression. The mechanism(s) through which Exisulind stimulates apoptosis in adenomatous tissue is unknown, however, *in vitro* studies suggest that Exisulind does not promote apoptosis through p53-dependent mechanisms.[17]

Mucin production in dysplastic glands was stimulated by treatment with Exisulind. This was observed in a high percentage of treated glands and was consistent with a return to a more normal glandular appearance. In association with the increased mucin, the nuclei became less hyperchromatic and the thickness of the glandular epithelium was reduced. This observation needs to be examined in other studies, including a determination of the nature of the mucin being produced. In addition, studies should be carried out to determine the underlying molecular mechanism(s) by which Exisulind stimulates mucin production in dysplastic glands.

In summary, treatment of FAP patients with sulindac sulfone for a period of six months led to a stabilization in the number of rectal polyps and a regression of smaller polyps. Polyp regression appeared to be caused by Exisulind-induced stimulation of adenoma cell differentiation and apoptosis. Exisulind does not appear to be as potent as sulindac in regressing polyps in FAP patients but, unlike sulindac, the drug does not appear to produce major gastrointestinal disturbances. Phase II clinical trials involving FAP patients are currently in progress.

REFERENCES

1. Bussey H.J.R., Eyers A.A., Ritchie S.M., and Thompson J.P.S. The rectum in adenomatous polyposis: the St. Mark's policy. *Br. J. Surg.* 72:S29–S35, 1985.
2. Burt R.W. and Gorden J. The genetic and molecular diagnosis of adenomatous polyposis coli. *Gastroenterology* 104:1211, 1993.
3. Rigas B., Goldman I.S., and Levine L. Altered eicosanoid levels in human colon cancer. *J. Lab. Clin. Med.* 122:518, 1993.
4. Waddell W.R., Ganser G.F., Cerise E.J., and Loughry R.W. Sulindac for polyposis of the colon. *Am. J. Surg.* 157:175, 1989.
5. Rigau J., Pique J.M., Rubio E., Planas R., Tarrech J.M., and Bordas J.M. Effects of long-term sulindac therapy on colonic polyposis. *Ann. Intern. Med.* 115:952, 1991.
6. Labayle D., Fischer D., Vielh P., et al. Sulindac causes regression of rectal polyps in familial adenomatous polyposis. *Gastroenterology* 101:635, 1991.
7. Giardiello F.M., Hamilton S.R., Krush A.J., Piantadosi S., Hylind L.M., and Celano P. Treatment of colonic and rectal adenomas with sulindac in familial adenomatous polyposis. *New Engl. J. Med.* 328: 1313, 1993.
8. Moorghen M., Ince P., Finney K.J., Sunter J.P., Appleton D., and Watson A.J. A protective effect of sulindac against chemically-induced primary colonic tumors in mice. *J. Pathol.* 156:341, 1988.
9. Skinner S.A., Penney A.G., and O'Brian P.E. Sulindac inhibits the rate of growth and appearance of colon tumors in the rat. *Arch. Surg.* 126:1094, 1991.
10. Shen T.Y., Ham E.A., Cirillo V.J., et al. Prostaglandin synthetase inhibitors. In: Robinson III and Vane J.R. (eds) Prostaglandin synthetase inhibitors, Their effects on physiological functions and pathological states. Raven Press, New York, 18, 1974.
11. Shen T.Y. and Winter C.A. Chemical and biological studies of indomethacin, sulindac, and their analogs. *Adv. Drug. Res.* 12:90, 1977.
12. Piazza G.A., Kulchak Rahm A.L., Krutzsch M., et al. Antineoplastic drugs sulindac sulfide and sulfone inhibit cell growth by inducing apoptosis. *Cancer Res.* 55:3110, 1995.
13. Piazza G.A., Alberts D.S., Hixson L.J., et al. Sulindac sulfone inhibits azoxymethane-induced colon carcinogenesis in rats without reducing prostaglandin levels. *Cancer Res.* 57(14):2909, 1997.
14. van Stolk R., Hayton W.L., Chan K.K., et al. Safety and tolerability of sulindac sulfone (Exisulind) in patients with familial adenomatous polyposis, results of a phase I and pharmacokinetic trial. *Gastroenterology*, submitted. 1998.
15. Cattoretti G., Becker M.H., Key G., et al. Monoclonal antibodies against recombinant parts of the ki-67 antigen (MIB 1 and MIB 3) detect proliferating cells in microwave-processed formalin-fixed sections. *J. Pathol.* 168:357, 1992.
16. Gavrieli Y., Sherman Y., and Ben-Sasson S.A. Identification of prolonged cell death *in situ* via specific labeling of nuclear DNA fragmentation. *J. Cell. Biol.* 119:493, 1992.
17. Piazza G.A., Rahm A.K., Finn T.S., et al. Apoptosis primarily accounts for the growth-inhibitory properties of sulindac metabolites and involves a mechanism that is independent of cyclooxygenase inhibition, cell cycle arrest, and p53 induction. *Cancer Res.* 57:2452, 1997.

6

PREVENTION OF COLON CANCER AND MODULATION OF ABERRANT CRYPT FOCI, CELL PROLIFERATION, AND APOPTOSIS BY RETINOIDS AND NSAIDs

Michael A. Pereira

Department of Pathology
Medical College of Ohio
HEB, Rm 200F
3055 Arlington Ave.
Toledo, Ohio 43614-5803

1. INTRODUCTION

Cancer is a multistage process involving multiple rare events with each suggesting a mutational or chromosomal alteration. At the different steps along the progression to cancer, the cells accumulate selective growth advantage that could be manifested by an increase in cell proliferation and/or a decrease in apoptosis and terminal differentiation. Promotion of the progression to cancer could therefore involve agents, growth factors, proto-oncogenes, etc., which enhance cell proliferation while decreasing apoptosis. Chemopreventive agents, on the other hand, would be agents that decrease cell proliferation while enhancing apoptosis resulting in a decrease in the selective growth advantage of precancerous lesions. A putative early precancerous lesion in the colon is the aberrant crypt focus (ACF) described and reviewed by Bird.[1] Increased cell proliferation appears to result in the accumulation of epithelial cells at the lumina of the crypts and hence the larger appearance of the aberrant crypts. In laboratory animals ACF are induced by chemical carcinogens and presumably contain precancerous cells that have many characteristics in common with colon tumors including in some dysphasia and an activated K-*ras* oncogene.[2–4] K-*ras* and APC mutations have also been demonstrated in ACF in human colon.[5–7] ACF are proposed to progress further to adenomas and finally cancer. During this progression, there is an increase in the level of cell proliferation with less of an increase in apoptosis, resulting in the overall growth of the lesions.

Colon Cancer Prevention, edited under the auspices of AICR
Kluwer Academic / Plenum Publishers, New York, 1999.

In the studies reported here, we used the azoxymethane (AOM)-induced colon cancer model to evaluate the ability of retinoids and nonsteroidal anti-inflammatory drugs (NSAIDs) to prevent colon cancer and ACF. The use of the AOM-induced colon cancer model to demonstrate chemoprevention and its modification to determine the ability of agents to prevent of ACF as a screen for chemopreventive agents have been reviewed.[8] The experimental design of the AOM model consisted of administering two doses of 15 mg/kg body weight of AOM, each one week apart. The agents being evaluated for chemoprevention were administered in the diet starting at either one week before or 12 weeks after the AOM and continuing until the rats were sacrificed. Two other proposed biomarkers for chemoprevention are the ability to decrease cell proliferation and to increase apoptosis.[9,10] Hence, their modulation by retinoids and NSAIDs was determined as part of the evaluation of the two biomarkers in screening for chemopreventive activity.

2. PREVENTION OF COLON CANCER BY RETINOIDS

We have evaluated the ability of retinoids to prevent both ACF and colon cancer induced by AOM.[11,12] Thirteen different retinoids were evaluated for the ability to prevent AOM-induced ACF. 2-(Carboxyphenyl) retinamide (2-CPR) was the most potent with 4-(hydroxyphenyl) retinamide (4-HPR) and 9-*cis*-retinoic acid being among the next most potent retinoids.[11] We also evaluated the ability of the retinoids to reduce the level of cell proliferation in both ACF and non-involved crypts of the colon mucosa. Two measurements of cell proliferation were used; one being the Proliferating Cell Nuclear Antigen (PCNA)-labeling index and the other, the PCNA compartment size in the crypts. The ability of the retinoids to prevent ACF correlated very well with their ability to decrease the PCNA-labeling index and the PCNA compartment in both ACF and non-involved crypts. The best correlation was between the prevention of ACF and the reduction in the PCNA-labeling index in the ACF. Hence, 2-CPR was also the most potent in decreasing cell proliferation (both PCNA-labeling index and proliferative compartment) with 4-HPR and 9-*cis*-retinoic acid again being among the next most potent retinoids. Thus, the ability of the retinoids to prevent ACF and to decrease cell proliferation appeared to be associated.

The ability of 2-CPR, 4-HPR, and 9-*cis*-retinoic acid to prevent AOM-induced colon cancer was then determined.[11,12] 4-HPR and 9-*cis*-retinoic acid were administered starting one week prior to the first of two weekly 15 mg/kg injections of AOM for a total of 36 weeks.[11] Both 4-HPR and 9-*cis*-retinoic acid produced 40–50% reduction in the yield of colon tumors similar to their reduction in the yield of ACF. 2-CPR was evaluated in another study by administering it starting either two weeks before or 12 weeks after the second dose of AOM.[12] The animals continued to receive 2-CPR until sacrificed 46 weeks after the first dose of AOM. Unexpectedly, 2-CPR administered starting either before or after AOM more than doubled the yield of colon tumors. The yields of both adenomas and adenocarcinomas were increased by 2-CPR with the shorter exposure starting after AOM increasing the yield of adenomas and the longer exposure starting before AOM increasing adenocarcinomas. Hence, the longer the exposure to 2-CPR the greater the stimulation of the progression to cancer. However, the enhancement of the total number of tumors was not related to the duration of exposure, being the same whether exposure started before or after AOM. Thus, 2-CPR has the characteristics of a tumor promoter, increasing the rate of progression to cancer, but not increasing the extent of initiation.

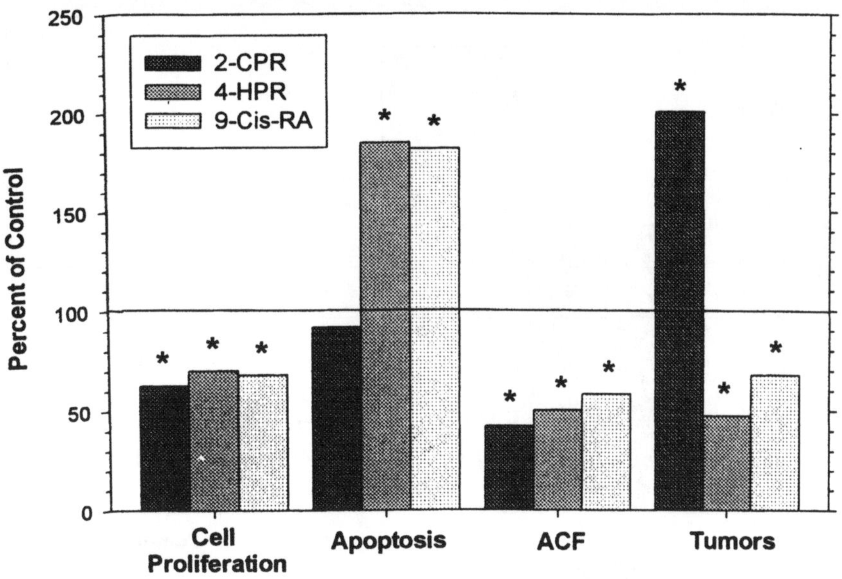

Figure 1. Summary of the effect of 2-CPR, 4-HPR, and 9-*cis* retinoic acid on cell proliferation, apoptosis, and AOM-induced ACF and tumors in the colon of F344 rats. The * indicates a significant difference (p-value < 0.05) from animals administered AOM and control diet. The results are from Zheng et al.[11,12]

The adenomas from animals exposed to the three retinoids were evaluated for treatment related effect on cell proliferation and apoptosis.[12] Cell proliferation in adenomas was reduced by all three retinoids similar to what was previously observed in ACF and non-involved crypts. Furthermore, the two retinoids, 4-HPR and 9-*cis*-retinoic acid that prevented colon tumors enhanced the level of apoptosis in the adenomas. In contrast, 2-CPR did not affect the level of apoptosis in adenomas. To determine whether the retinoids induced these effects on cell proliferation and apoptosis in adenomas in contrast to selection of adenomas with the characteristic, animals with AOM-induced colon tumors were exposed to 4-HPR or 2-CPR for only six days prior to sacrifice. 4-HPR decreased cell proliferation and enhanced apoptosis while 2-CPR only reduced cell proliferation in adenomas and non-involved crypts, indicating that the retinoids induced the effects on cell proliferation and apoptosis in adenomas rather than in selected adenomas with the characteristics.

Figure 1 summarizes our results with 2-CPR, 4-HPR, and 9-*cis* retinoic acid. All three retinoids were similarly potent in preventing ACF and reducing cell proliferation. However, only the two retinoids that prevented cancer enhanced the level of apoptosis. Furthermore 2-CPR strongly promoted instead of preventing colon cancer.

3. PREVENTION OF COLON CANCER BY NSAIDs

3.1. Piroxicam

Piroxicam is very potent in preventing AOM-induced colon tumors and ACF when administered starting either one week before or 12–14 weeks after AOM.[13–17] Furthermore, treatment with piroxicam caused the regression of AOM-induced ACF that

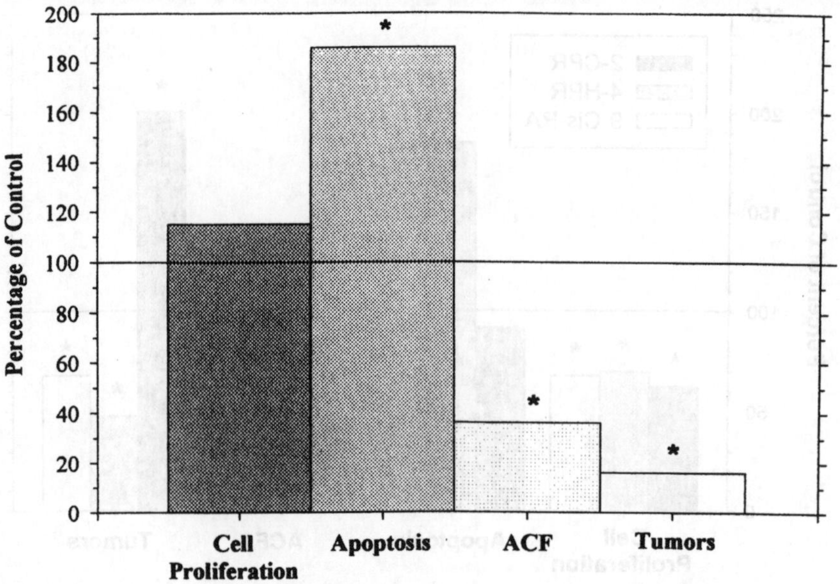

Figure 2. Effect of piroxicam on cell proliferation, apoptosis, and AOM-induced ACF and colon tumors. The * indicates a significant difference from animals administered AOM + control diet (p-value < 0.05). The ACF and tumor data are from Pereira et al.[12] and the cell proliferation and apoptosis results are unpublished.

reoccurred upon termination of treatment.[17] Our unpublished results demonstrate that when piroxicam was administered 12 weeks after the second dose of AOM, the level of ACF was reduced from 168 ± 25 to 73 ± 12 within one week of treatment. When treatment with piroxicam was terminated 7 or 16 weeks later, the yield of ACF increased to 115 ± 15 and 132 ± 15, respectively which was not different from the yield of 122 ± 16 in animals administered AOM + control diet. Furthermore, when treatment was terminated after 7 and 16 weeks, the yield of tumors/animal at week 46 was 2.65 ± 0.35 and 2.58 ± 0.37, respectively, also not significantly different from controls (3.21 ± 0.28). However, continuous treatment with piroxicam from week 11 to 46 after AOM significantly reduced the yield of colon tumors/animal from 3.21 ± 0.28 to 0.50 ± 0.19. Although piroxicam was very efficient in preventing colon tumors, it did not affect the level of cell proliferation in the adenomas, but it increased the level of apoptosis so long as it was administered until the animals were sacrificed. Hence, continued treatment with piroxicam appears to be required to maintain an enhanced level of apoptosis and to prevent AOM-induced ACF and tumors. Figure 2 summarizes the results of these experiments that indicate that piroxicam is very potent in preventing AOM-induced ACF and colon tumors and in enhancing apoptosis while not affecting cell proliferation in the adenomas.

3.2. Aspirin

Aspirin has been reported to prevent AOM-induced ACF and colon cancer in rats.[18–20] Dose response studies were performed to compare the effect of aspirin on cell proliferation, apoptosis, and AOM-induced ACF and colon tumors (Fig. 3).[20] Aspirin at 600 and 1800 mg/kg diet administered starting before AOM was very potent in reduc-

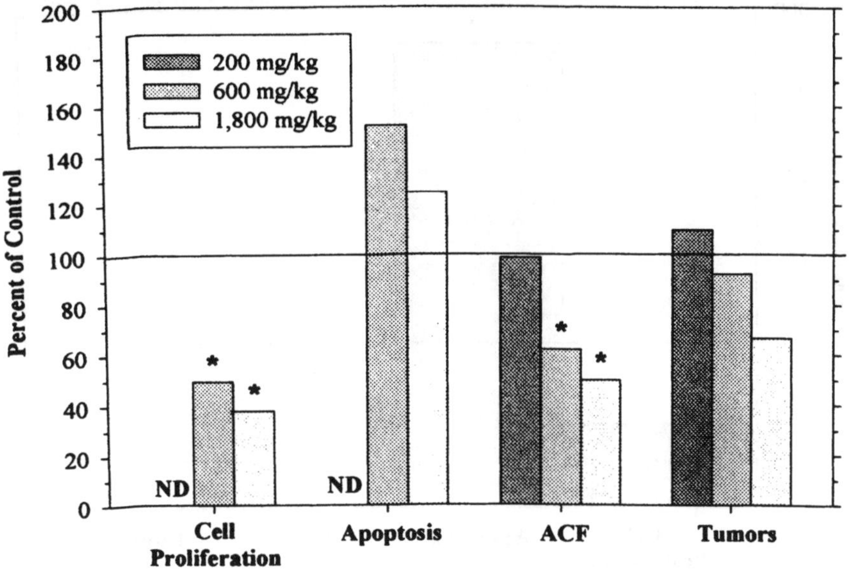

Figure 3. Effect of aspirin on cell proliferation, apoptosis, and AOM-induced ACF and colon tumors. The * indicates a significant difference from animals administered AOM + control diet (p-value < 0.05) and ND indicates that the effect of the 200 mg/kg dose of aspirin was not determined. The results are from Li et al.[20]

ing cell proliferation and preventing ACF. However, neither concentration of aspirin significantly affected the level of apoptosis or the yield of AOM-induced colon tumors. When aspirin was administered at 600 and 1800 mg/kg diet starting after AOM, the tumor yield was respectively, 100 and 53% (p-value < 0.05) of the yield in the AOM + control diet group. This indicated that a very high dose of aspirin was minimally effective in preventing colon cancer while prevention of ACF and reduction of cell proliferation were more susceptible at lower concentrations. The low efficacy of aspirin in preventing colon cancer could be related to its inability to significantly enhance the level of apoptosis. The prevention of colon cancer would then be the result of the decrease in cell proliferation.

3.3. α-Difluoromethylornithine (DFMO)

DFMO is a strong inhibitor of ornithine decarboxylase, the rate limiting enzyme in polyamine biosynthesis required for cell proliferation.[21-22] In laboratory animal models, DFMO has prevented cancer including AOM-induced colon cancer.[14-16] We have evaluated the effect of DFMO on cell proliferation, apoptosis and AOM-induced ACF, and colon tumors (Fig. 4).[20] When administered starting 12 weeks after the last dose of AOM, DFMO at 1000 and 3000 mg/kg diet was very potent in preventing colon cancer, reducing the yield by approximately 70%. In adenomas that occurred in the presence of DFMO, the level of cell proliferation was reduced while the level of apoptosis was enhanced. When administered starting one week prior to AOM, DFMO (4000 mg/kg diet) reduced the yield of AOM-induced ACF.[23] Hence, DFMO reduced the level of cell proliferation, ACF, and colon tumors while enhancing the level of apoptosis in the tumors.

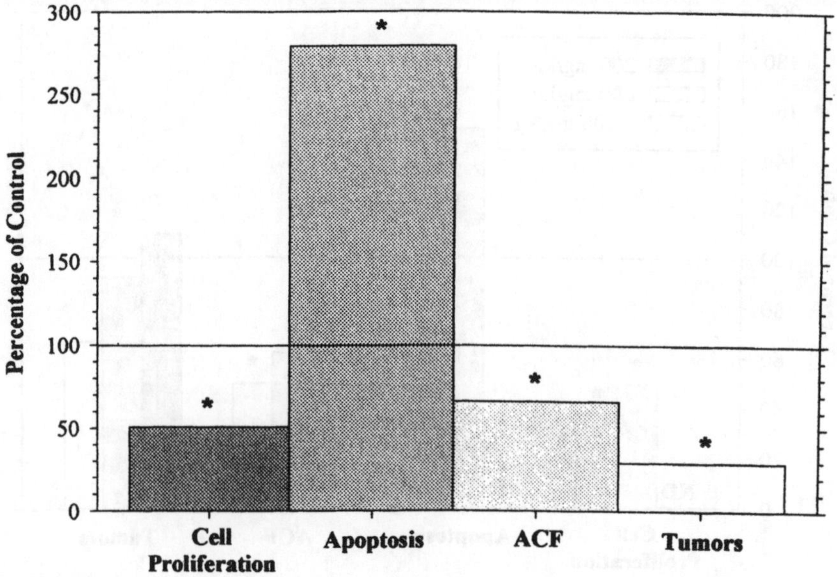

Figure 4. Effect of DFMO on cell proliferation, apoptosis, AOM-induced ACF, and colon tumors. The * indicates a significant difference from animals administered AOM + control diet (p-value < 0.05). The cell proliferation, apoptosis, and colon tumor data are from Li et al.[20] and the ACF data from Pereira and Khoury.[23]

4. SUMMARY AND CONCLUSIONS

The effect of the NSAIDs, retinoids and DFMO on AOM-induced colon tumors, and ACF, cell proliferation, and apoptosis is summarized in Table 1. The ability to prevent AOM-induced ACF has been used as an assay to screen agents for chemoprevention.[8] As discussed above, all six potential chemopreventive agents, aspirin, 2-CPR, DFMO, 4-HPR, piroxicam, and 9-*cis*-retinoic acid, decreased the level of AOM-induced ACF. However, two of the agents, aspirin (at doses that greatly reduced the yield of ACF) and 2-CPR did not prevent AOM-induced colon tumors. Hence, aspirin and 2-CPR would appear to be false positive in the ACF assay. Besides being a false positive in the ACF assay, 2-CPR actually had the opposite effect of doubling the yield of colon tumor. The false positive result for aspirin could be due to the lower sensitivity of the AOM-induced colon cancer assay compared to the ACF assay. However, aspirin

Table 1. Summary of effects of potential chemopreventive agents

Agent	Tumors	ACF	Cell Proliferation	Apoptosis
1. Aspirin	No Effect	Decreased	Decreased	No Effect
2. 2-CPR	Increased	Decreased	Decreased	No Effect
3. DFMO	Decreased	Decreased	Decreased	Increased
4. 4-HPR	Decreased	Decreased	Decreased	Increased
5. Piroxicam	Decreased	Decreased	No Effect	Increased
6. 9-Cis-RA	Decreased	Decreased	Decreased	ND

ND indicates not determined.

significantly reduced the yield of ACF at a dose (600 mg/kg diet) one-third the dose (1800 mg/kg diet) that did not reproducibly reduce the yield of colon tumors. Thus, although there were no false negative results, two of the six agents gave false positive results in the AOM-induced ACF assay with respect to their ability to prevent colon cancer.

Two other potential biomarkers for chemopreventive activity are the ability to reduce the level of cell proliferation and to enhance the level of apoptosis. All six of the agents including aspirin and 2-CPR reduced the level of cell proliferation in adenomas. Thus, similar to their ability to prevent ACF, the ability of aspirin and 2-CPR to decrease cell proliferation were also false positive responses with respect to prevention of colon cancer, but not with respect to the prevention of ACF. Piroxicam, the most potent of the six agents in preventing AOM-induced colon cancer, did not significantly affect the level of cell proliferation in adenomas which is a false negative response. Hence, only three of the six agents (50%) were correctly identified as potential chemopreventive agents by their ability to reduce the level of cell proliferation. In contrast, retinoids, including the three discussed here, demonstrated good correlation between the ability to prevent AOM-induced ACF and the ability to decrease cell proliferation in colonic mucosa or ACF.[11] Thus, within some classes of agents such as the retinoids, the ability to prevent ACF and to reduce cell proliferation appear to correlate, while in other classes including the NSAIDs, the correlation appeared not to exist.

The four agents that prevented colon cancer all enhanced the level of apoptosis, while the two agents that did not prevent colon cancer did not effect apoptosis. Three other chemopreventive agents, including phenylethyl-3-methylcaffeate and the NSAIDs, curcumin and sulindac, have been shown by Samaha et al.[24] to enhance apoptosis in AOM-induced colon tumors. Thus, although a very limited number of chemopreventive agents have been evaluated for the ability to enhance apoptosis in the colon, there appears to be an association between the ability to enhance apoptosis and the ability to prevent colon cancer.

The use of the AOM-induced ACF assay to screen agents for the ability to prevent colon tumors would appear to result in false positive responses including agents (2-CPR and quercetin[25]) that actually promote colon cancer. However, our results suggest that false positive responders could be distinguished by their inability to enhance apoptosis while potential chemopreventive agents would enhance it. It is therefore proposed that a Two Step Procedure be used to screen agents for the ability to prevent colon cancer. Step 1 would be the determination of the ability to prevent ACF. Because the ACF assay appears to suffer more from false positive than from false negative responders, apparently few potent chemopreventive agents would be missed. Also the ACF assay could be the source of foci for evaluation of the effect of the agents on apoptosis. Thus, after evaluating the colon for ACF, tissue sections would be obtained, stained with hematoxylin and eosine and evaluated as Step 2 for enhancement of apoptosis in ACF and non-involved crypts. However, more potential chemopreventive agents, especially of different mechanisms and classes than retinoids and NSAIDs, need to be evaluated for the ability to prevent ACF and to enhance apoptosis in order to determine how general the conclusions presented are for the different types of agents. Furthermore, the correlation between the response in the Two Step Procedure and the ability to prevent colon cancer needs to be determined in order to validate the Two Step Procedure for screening agents for chemoprevention of colon cancer.

ACKNOWLEDGMENTS

Work reported from our laboratory was supported in part by the National Cancer Institute NO1-CN-55151, NO1-CN-55174, and NO1-CN-55175.

REFERENCES

1. Bird, R.P. Role of aberrant crypt foci in understanding the pathogenesis of colon cancer. *Cancer Lett.* 93:55–71, 1995.
2. Stopera, S.A., Murphy, L.C., and Bird, R.P. Evidence for a *ras* gene mutation in azoxymethane-induced colonic aberrant crypts in Sprague-Dawley rats: earliest recognizable precursor lesions of colon cancer. *Carcinogenesis* 13:2081–2085, 1992.
3. Vivona, A.A., Shpitz, B., Medline, A., Bruce, W.R., Hay, K., Ward, M.A., Stern, H.S., and Gallinger, S. K-*ras* mutations in aberrant crypt foci, adenomas, and adenocarcinomas during azoxymethane-induced colon carcinogenesis. *Carcinogenesis* 14:1777–1781, 1993.
4. Shivapurkar, N., Tang, Z., Ferreira, A., Nasim, S., Garett, C., and Alabaster, O. Sequential analysis of K-*ras* mutations in aberrant crypt foci and colonic tumors induced by azoxymethane in Fischer-344 rats on high-risk diet. *Carcinogenesis* 15:775–778, 1994.
5. Smith, A.J., Stern, H.S., Penner, M., Hay, K., Itri, A., Bapat, B.V., and Gallinger, S. Somatic *APC* and K-*ras* codon 12 mutations in aberrant crypt foci from human colons. *Cancer Res.* 54:5527–5530, 1994.
6. Pretlow, T.P., Brasitus, T.A., Fulton, N.C., Cheyer, C., and Kaplan, E.L. K-*ras* mutations in putative preneoplastic lesions in human colon. *J. Natl. Cancer Inst.* 85:2004–2007, 1993.
7. Jen, J., Powell, S.M., Papadopoulos, N., Smith, K.J., Hamilton, S.R., Vogelstein, B., and Kinzler, K.W. Molecular determinants of dysplasia in colorectal lesions. *Cancer Res.* 54:5523–5526, 1994.
8. Steele, V.E., Moon, R.C., Lubet, R.A., Grubbs, C.J., Reddy, B.S., Wargovich, M., McCormick, D.L., Pereira, M.A., Crowell, J.A., Bagheri, D., Sigman, C.C., Boone, C.W., and Kelloff, G.J. Preclinical efficacy evaluation of potential chemopreventive agents in animal carcinogenesis models: Methods and results from the NCI chemoprevention drug development program. *J. Cellular Biochem.*, Suppl. 20:32–54, 1994.
9. Vogelstein, B. and Kinzler, K.W. The multistep nature of cancer. *Trends Genetics* 9:138–141, 1993.
10. Kelloff, G.J., Boone, C.W., Crowell, J.A., Steele, V.E., Lubet, R., and Sigman, C.C. Chemopreventive drug development: Perspectives and progress. *Cancer Epidemiol. Biomarkers Prev.* 3:85–98, 1994.
11. Zheng, Y., Kramer, P.M., Olson, G., Lubet, R.A., Steele, V.E., Kelloff, G.J., and Pereira, M.A. Prevention by retinoids of azoxymethane-induced colon cancer and aberrant crypt foci and their modulation of cell proliferation in a colon of rats. *Carcinogenesis* 18:2119–2125, 1997.
12. Zheng, Y., Kramer, P.M., Lubet, R.A., Steele, V.E., Kelloff, G.J., and Pereira, M.A. Effect of retinoids on AOM-induced colon cancer in rats: Modulation of cell proliferation, apoptosis and aberrant crypt foci. *Carcinogenesis*. In press, 1998.
13. Reddy, B.S., Maruyama, H., and Kelloff, G. Dose-related inhibition of colon carcinogenesis by dietary piroxicam, a nonsteroidal anti-inflammatory drug, during different stages of rat colon tumor development. *Cancer Res.* 47:5340–5346, 1987.
14. Reddy, B.S., Nayini, J., Tokumo, K., Rigotty, J., Zang, E., and Kelloff, G. Chemoprevention of colon carcinogenesis by concurrent administration of piroxicam, a nonsteroidal anti-inflammatory drug with DL-α-difluoromethylornithine, an ornithine decarboxylase inhibitor, in diet. *Cancer Res.* 50:2562–2568, 1990.
15. Rao, C.V., Tokumo, K., Rigotty, J., Zang, E., Kelloff, G., and Reddy, B.S. Chemoprevention of colon carcinogenesis by dietary administration of piroxicam, α-difluoromethylornithine, 16α-fluoro-5-androsterone, and ellagic acid individually and in combination. *Cancer Res.* 51:4528–4534, 1991.
16. Kulkari, N., Zang, E., Kelloff, G., and Reddy, B.S. Effect of the chemopreventive agents piroxicam and D,L-α-difluoromethylornithine on intermediate biomarkers of colon carcinogenesis. *Carcinogenesis* 13:995–1000, 1992.
17. Pereira, M.A., Barnes, L.H., Steele, V.E., Kelloff, G.J., and Lubet, R.A. Piroxicam-induced regression of azoxymethane-induced aberrant crypt foci and prevention of colon cancer in rats. *Carcinogenesis* 17:373–376, 1996.

18. Wargovich, M.J., Chen, C.D., Harris, C., Yang, E., and Velasco, M. Inhibition of aberrant crypt growth by non-steroidal anti-inflammatory agents and differentiation agents in the rat colon. *Int. J. Cancer* 60:515–519, 1995.
19. Reddy, B.S., Rao, C.V., Rivenson, A., and Kelloff, G.J. Inhibitory effect of aspirin on azoxymethane-induced colon carcinogenesis in F344 rats. *Carcinogenesis* 14:1493–1497, 1993.
20. Li, H., Schut, H.A.J., Conran, P., Kramer, P.M., Lubet, R.A., Steele, V.E., Kelloff, G.J., and Pereira, M.A. Prevention by aspirin and its combination with difluoromethyl-ornithine of azoxymethane-induced tumors and aberrant crypt foci in the colon of rats. *Carcinogenesis*. In press, 1998.
21. Pegg, A.E. Recent advances in the biochemistry of polyamines in eukaryotes. *Biochem. J.* 234:249–262, 1986.
22. McCann, P.P., Bitoni, A.J., and Pegg, A.E. Inhibition of polyamine metabolism and the consequent effects on cell proliferation, *In* Wattenberg, L., Lipkin, M., Boone, C.W., and Kelloff, G.J. (eds), Cancer Chemoprevention. CRC Press, Boca Raton, FL, pp. 531–539, 1992.
23. Pereira, M.A. and Khoury, M.D. Prevention by chemopreventive agents of azoxymethane-induced foci of aberrant crypts in rat colon. *Cancer Lett.* 61:27–33, 1991.
24. Samaha, H.S., Kelloff, G.J., Steele, V., Rao, C.V., and Reddy, B.S. Modulation of apoptosis by sulindac, curcumin, phenylethyl-3-methylcaffeate, and 6-phenylhexyl isothiocyanate: Apoptotic index as a biomarker in colon cancer chemoprevention and promotion. *Cancer Res.* 57:1301–1305, 1997.

7

DIGESTED FIBER FROM WHEAT BRAN INDUCES cdk INHIBITORS WHICH BLOCK COLON EPITHELIAL CELLS IN G1

Eileen A. Friedman and Jingping Wang

Department of Pathology
State University of New York Health Science Center at Syracuse
Syracuse, New York 13210

1. INTRODUCTION

We have tested the hypothesis that short chain organic acids in the colon derived from dietary pectin, wheat bran, oat bran, and cellulose are protective against the development of colon cancer because they cause growth arrest by inducing cyclin-directed kinase (cdk) inhibitors. A more quiescent colonic epithelium is less susceptible to mutagens which can lead to mutations or deletions in the major genes implicated in the progression to colon cancer: ras, p53, and APC[1] or in the genes involved in TGFβ response such as TβRII[2] or the Smad genes which are intermediates in TGFβ signaling.[3,4] TGFβ1 is a small protein found *in vivo* which is the most potent epithelial cell growth inhibitor known. High levels of TGFβ1 mRNA and peptide are found at the top of the colonic crypt in the region of terminally differentiated cells.[5,6] Injected TGFβ1 shortens colonic crypts.[7] These observations suggest that TGFβ1 induces differentiation and growth arrest *in vivo.* TGFβ1 inhibits colonocyte proliferation by inhibiting phosphorylation of the retinoblastoma protein through blocking the activity of cdk/cyclin complexes. TGFβ1 inhibits cdk kinases by inducing the cdk inhibitors p15 and p21 and causing accumulation of p27.[8–10] Thus short chain organic acids in the colon derived from dietary fiber could act directly to induce cdk inhibitors, or they could work indirectly by first inducing TGFβ1, which in turn induced cdk inhibitors.

These hypotheses were tested using the U4 human colon carcinoma cell line which is capable of differentiation into a cell modeling the major colonocyte lineage found in the colon, the fluid-transporting enterocyte. U4 cells differentiate when treated with agents that upregulate expression of TGFβ1 mRNA and biologically active mature TGFβ1. TGFβ1 induces differentiation markers in U4 cells and inhibits U4 cell growth by blocking cells in G1.[11]

Colon Cancer Prevention, edited under the auspices of AICR
Kluwer Academic / Plenum Publishers, New York, 1999.

Table 1. Mixtures of short chain fatty acids mimicking fiber digestion products

Fiber	Butyrate	Propionate	Acetate
wheat bran	4.7 mM	3.3 mM	25 mM
oat bran	2.7 mM	3.3 mM	27 mM
pectin	1.0 mM	15 mM	30 mM
cellulose	0.5 mM	2.0 mM	13 mM

2. EXPERIMENTAL

2.1. SCFAs Mimicking Digested Fiber Induce Growth Arrest Like TGFβ1

We made mixtures of SCFA which would mimic the colonic SCFA content of rats fed synthetic diets with the fiber provided by either wheat bran, oat bran, pectin or cellulose, summarized in Table 1.[12–15] The authors cited demonstrated that a wheat bran diet led to colonic contents rich in butyrate with less propionate. The oat bran diet yielded about half as much butyrate and a similar amount of propionate as the wheat bran diet. In contrast, the pectin diet yielded a SCFA mileu primarily rich in propionate while the cellulose was largely unfermentable and led to lower levels of all SCFA with only 10% as much butyrate as the wheat bran diet.

We then tested the effects of the SCFA mixtures *in vitro* in serum-free chemically defined media on U4 human colon carcinoma cells which are capable of differentiation into transporting enterocytic cells.[11] Treatment of asynchronous U4 pre-enterocytic cells with the SCFA mixtures for 24 hours led to alterations in the cell cycle. The wheat bran SCFAs caused a 60% decrease in the percentage of U4 cells in S phase as tested by cytofluorometry with the cells arresting in G1, while the SCFA mixtures corresponding to either oat bran, pectin or cellulose had either marginal or undetectable effects. After 48 hours of exposure, growth inhibitory effects of the other fiber SCFA mixtures became evident by cytofluorometry. Supporting these observations, after 48 hours of treatment with wheat bran, oat bran, pectin or cellulose fiber SCFA mixtures, a 3 hour pulse label of ^{3}H-TdR showed decreases in labeling of 89+/–2%, 84+/–3%, 79+/–1%, and 38+/–5%, respectively. The effects of the SCFA mixtures after 48 hours on growth arrest in G1 by cytofluorometry were averaged with the effects of these agents on inhibition of ^{3}H-TdR labeling. Wheat bran, oat bran, pectin or cellulose fiber SCFA mixtures induced, respectively, decreases in the number of cycling cells of 75%, 42%, 40%, and 19% (Fig. 1). Thus, of the fiber SCFA mixtures tested, the major effect on cell proliferation was elicited by the wheat bran fiber SCFAs, and fiber SCFAs arrested U4 cells in G1.

2.2. Fiber SCFAs Induce cdk Inhibitors p21 and p27

Cell cycle entry into S phase is controlled by the complexing of an inactive kinase or cdk (cyclin-directed kinase) to an activating cyclin. The activated cdk/cyclin kinase complex then phosphorylates the retinoblastoma protein, releasing its trapped transcription factors which mediate entry into S phase. The cdk4/cyclin D and cdk2/cyclin

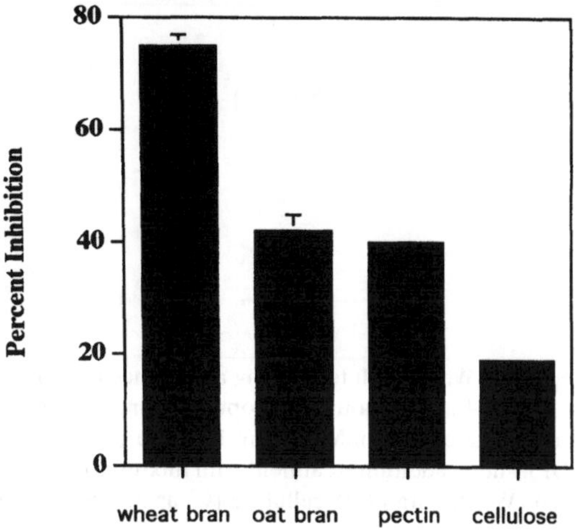

Figure 1. Inhibition of cell cycling by digested fiber SCFAs mean of cytofluography & ^{3}H-TdR data. Mean of cell cycle analysis of asynchronous U4 cells by cytofluorography and ^{3}H-TdR incorporation after 48 hours of treatment with a mixture of SCFAs mimicking wheat bran fiber, oat bran fiber, pectin, or cellulose digestion products found in the colon. A mean of 7634 +/− 739 (SE) cells was analyzed for each point by cytofluorography while a mean of 2×10^6 cells was analyzed for ^{3}H-TdR incorporation. Error bars less than 5% of the mean are not placed on the figure in the program used.

E complexes can phosphorylate the retinoblastoma protein to allow this transition into S phase, but the activity of these cdks, and thus cell cycle progression, can be blocked by cdk inhibitors. These inhibitors fall into two categories of structurally related molecules: p21, p27, and p57, which inhibit multiple cdk/cyclin complexes,[10,16–18] and p15, p16, p18, and p19 which interact directly with cdk4 and cdk6 and inhibit dimerization with cyclins.[8,19] p16, p18, and p19 are not expressed on the protein level in U4 cells while p15 was not readily detectable by western blotting (data not shown). p57 expression is restricted to skeletal muscle, heart, brain, kidney, and lung in human tissue.[20] Therefore, we concentrated our studies on p21 and p27 which inhibit multiple cdk/cyclin complexes. We also assayed levels of cyclin D1 and cyclin E by western blotting as the levels of these cyclins determine whether cdk4 and cdk2 become activated. In addition to association with cyclins, activation of cdks requires phosphorylation of a threonine residue. Since this phosphorylation is carried out by the cdk7/cyclin H complex,[21] the abundance of cyclin H was also assayed after fiber SCFA treatment.

Treatment of U4 pre-enterocytic cells with SCFA mixtures mimicking digested wheat bran fiber (W) led to a large increase in the cdk inhibitor p21 which was maximal by 24 hours and maintained over control levels for 72 hours as detected by western blotting (Fig. 2, one of 4 duplicate experiments shown; Fig. 3). Oat bran (O) and pectin fiber (PC) SCFAs led to a smaller induction of p21. All of these increases were statistically significant at the 24 hour point. In contrast, the small increase induced by cellulose (C) fiber SCFA mixture was not statistically significant compared to untreated control (CT) cells. The SCFAs in the fiber mixtures were also tested individually at their optimal concentration for inducing U4 cell growth arrest, which had been determined by our earlier experiments (data not shown): 5 mM sodium butyrate (B), 15 mM sodium propionate (P), and 40 mM ammonium acetate (A). The individually applied SCFAs did not induce p21 to the same extent as the wheat bran fiber mixture, while the mixture of the optimal concentration (OP) of each SCFA after 24 and 72 hours of treatment was an average of 1.5 fold more effective than the wheat bran fiber SCFA mixture, as measured by scanning densitometry of the immunoblots (Fig. 2). Both the wheat bran and optimal SCFA mixture kept p21 levels elevated over control levels for 72 hours. Thus of all of the fiber SCFA mixtures tested, the wheat bran fiber

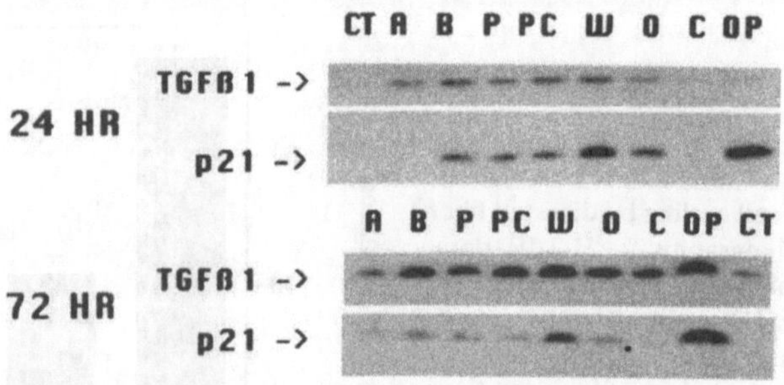

Figure 2. Western blots showing abundance of TGFβ1 and the cdk inhibitor p21cip1 after treatment of U4 cells for 24 and 72 hours with optimal concentrations of SCFAs for inducing growth arrest: 40mM ammonium acetate (A), 5mM sodium butyrate (B), 15mM sodium propionate (P), a mixture of these 3 SCFA (OP); and in addition treatment with mixtures of SCFAs mimicking digestion products of pectin (PC), wheat bran (W), oat bran (O), cellulose (C) as described in Table 1. CT, control untreated cells.

mixture came closest to an optimal combination of three SCFAs in inducing the cdk inhibitor p21.

Levels of the cdk inhibitor p27 and the cyclins D1, E, and H, were then assayed by western blotting after 24 and 48 hours exposure to the SCFA fiber mixtures. The SCFA mixture mimicking digested wheat bran fiber induced the largest increases in cdk inhibitors, with statistically significant inductions of p21 (30-fold) and of p27 (5-fold) (Figs. 3–7). These increases were maintained during 48 hours of treatment (data not shown). The small increases in cyclin D1 and cyclin H were not significant (data not shown). However, statistically significant increases in cyclin E levels by the pectin and wheat bran SCFA mixes were observed (Figs. 3&5). Therefore, fiber SCFA mixtures primarily increased abundance of the cdk inhibitors p21 and p27 and of cyclin E, but not the abundance of cyclins D1 or H. The induction of p21cip1 was greatest by wheat bran fiber > oat bran fiber > pectin > cellulose (Fig. 7). However, oat bran fiber and pectin induced almost the same growth inhibition (Fig. 1). Other factors induced by digested fiber must control colonocyte cycling in addition to p21cip1.

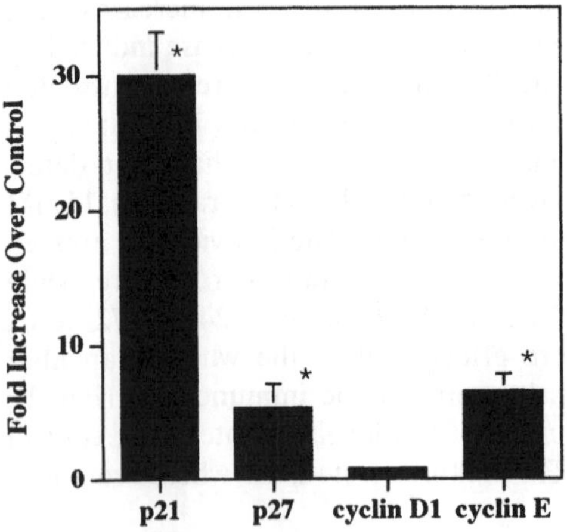

Figure 3. Induction of cdk inhibitors & cyclins by wheat bran fiber SCFA mix. Graphic depiction of increase in cdk inhibitors p21 and p27 and cyclin D1 and cyclin E after 48 hours of treatment of U4 cells with a SCFA mixture mimicking digested wheat bran fiber. Values statistically different from control by Student's t test indicated with asterisk.

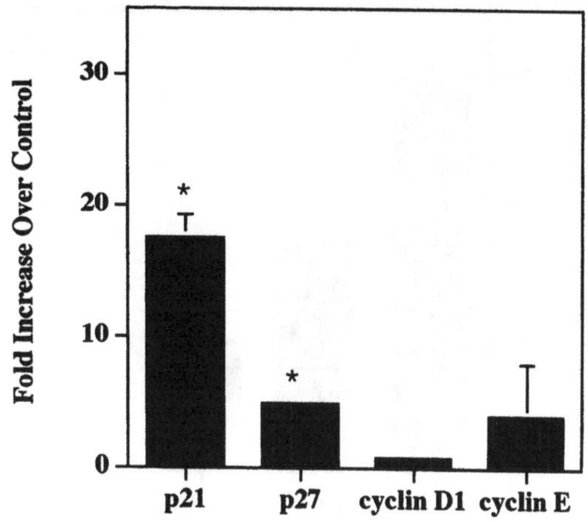

Figure 4. Induction of cdk inhibitors & cyclins by oat bran fiber SCFA mix. Graphic depiction of increase in cdk inhibitors p21 and p27 and cyclin D1 and cyclin E after 48 hours of treatment of U4 cells with a SCFA mixture mimicking digested oat bran fiber. Values statistically different from control by Student's t test indicated with asterisk.

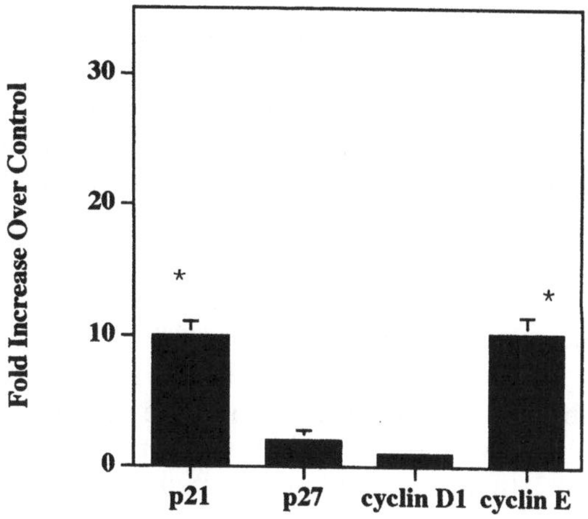

Figure 5. Induction of cdk inhibitors & cyclin E by pectin fiber SCFA mix. Graphic depiction of increase in cdk inhibitors p21 and p27 and cyclin D1 and cyclin E after 48 hours of treatment of U4 cells with a SCFA mixture mimicking digested pectin. Values statistically different from control by Student's t test indicated with asterisk.

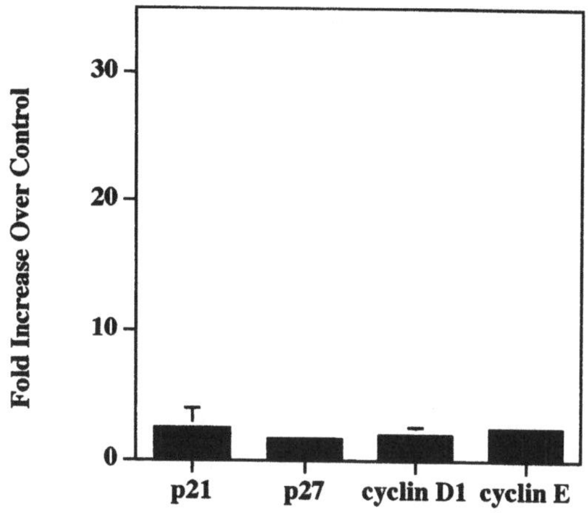

Figure 6. Lack of induction of cdk inhibitors & cyclins by cellulose fiber SCFA mix. Graphic depiction of lack of increase in cdk inhibitors p21 and p27 and cyclin D1 and cyclin E after 48 hours of treatment of U4 cells with a SCFA mixture mimicking digested cellulose.

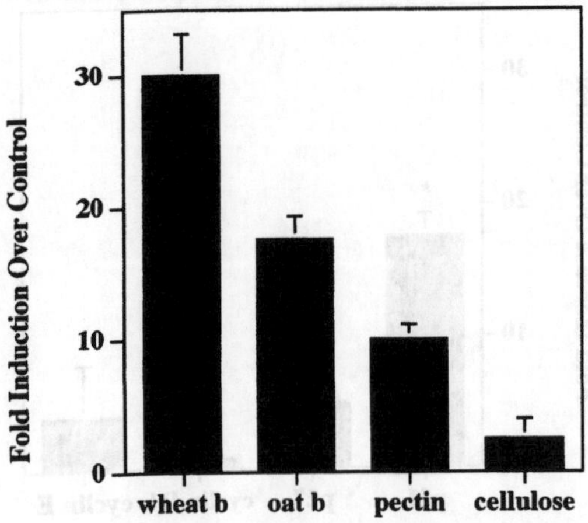

Figure 7. Induction of the cdk inhibitor p21cip1 by SCFAs mimicking fiber digests. Comparison of increase in the cdk inhibitor p21 after 48 hours of treatment of U4 cells with SCFA mixtures mimicking digested fibers.

2.3. Effect of SCFA Mixtures on cdk Kinase Activity

Cyclins activate cdks by complexing with them, while cdk inhibitors bind to cyclin/cdk complexes and inhibit their activity by blocking ATP binding sites. Were the increases in cyclin E abundance induced by wheat bran and pectin fiber SCFA mixtures capable of increasing cdk2 kinase activity, given the larger increases in cdk inhibitor concentration? To answer this question, we next measured the activities of cdk2 kinase and of cdk4 kinase and the abundance of cyclins, cdks, and cdk inhibitors in each cdk/cyclin complex.

U4 cells were treated with the SCFA mixture mimicking digested wheat bran fiber for 24 and 48 hours or were grown in parallel cultures but left untreated. Cdk2 activity was assayed by immunoprecipitating the cdk2/cyclin complex with antisera to cdk2, then assaying kinase activity on added histone H1. The phosphorylated histone H1 substrate was detected by autoradiography after SDS-PAGE. The wheat bran fiber SCFA mixture significantly inhibited cdk2 kinase activity after 24 and 48 hours of treatment (Fig. 8A), while 5 mM butyrate, one of the constitutents of the SCFA mixture, was slightly less effective. Analysis of the composition of the immunoprecipitated cdk2/cyclin complexes was performed by western blotting. At constant cdk2 levels (Fig. 8E), high levels of the cdk inhibitors p21 (Fig. 8B) and p27 (Fig. 8C) were found associated with the cdk2/cyclin complex, contributing to its decreased kinase activity. The greater inhibition of cdk2 kinase activity seen after 48 hours of treatment with wheat bran fiber SCFAs may reflect the increased amount of p27 associated with this complex at 48 versus 24 hours. Cdk inhibitors have been shown to cooperate to inhibit cdk2 kinase and cdk4 kinase activities.[10] While wheat bran fiber SCFA increased cytosolic cyclin E levels 6-fold (Fig. 3), a 6-fold increase in cyclin E associated with cdk2 was not seen in immunoblots of the cdk2 immunoprecipitates (Fig. 8D). Thus, the cdk2/cyclin E complexes bound the cdk inhibitors p21 and p27, which blocked cdk2 kinase activity (Fig. 8A). The increase in cytosolic cyclin E levels was not reflected in increased cdk2 kinase activity.

The abundance of cyclin D1 associated with cdk2 was not assayed because cdk2/cyclin D1 complexes are inactive kinases. Cdk2 associated with cyclin D1 is not phosphorylated by cdk7-cyclin H.[22]

Equal amounts of cdk4 (Fig. 9E) were immunoprecipitated from wheat bran

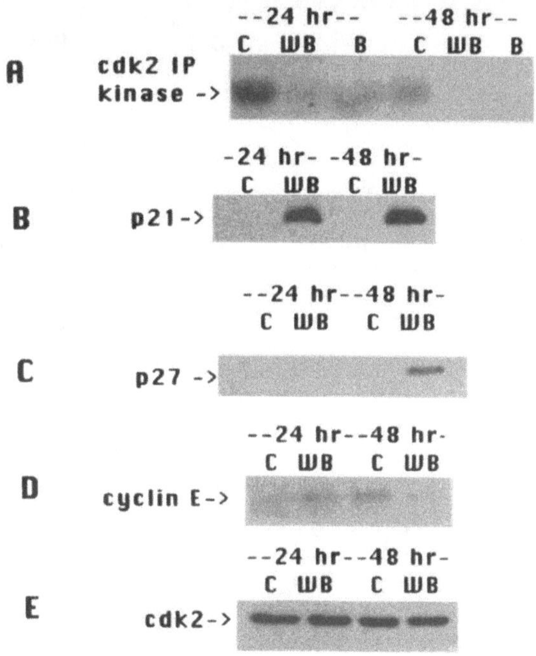

Figure 8. Analysis of cdk2/cyclin complexes immunoprecipitated from U4 cells which had been treated for 24 and 48 hours with either a mixture of SCFAs mimicking digestion products of wheat bran (WB) or with 5 mM butyrate (B); decreased kinase activity shown due to increased abundance of cdk inhibitors in the cdk2/cyclin E complex. **A.** immune complex kinase reaction using histone H1 as substrate; **B–E**, western blots of immunoprecipitated cdk2/cyclin complexes for the cdk inhibitor p21cip1 (**B**), the cdk inhibitor p27kip1 (**C**), cyclin E (**D**), and cdk2 (**E**) the control for equal immunoprecipitations.

SCFA mixture-treated and parallel untreated U4 cell cultures. Cdk4 kinase activity on histone H1 was assayed and the composition of the cdk4 immunoprecipitates was determined by immunoblotting. High levels of the cdk inhibitor p21 (Fig. 9B) were found associated with the cdk4/cyclin D1 complexes from wheat bran SCFA treated cells, contributing to the decreased kinase activity of the cdk4/cyclin D1 complexes (Fig. 9A). The greater abundance of p21 associated with cdk4 after 48 hours of treatment was reflected in a greater inhibition of cdk4 kinase activity at this time point. Increased amounts of the cdk inhibitor p27 were also found associated with cdk4/cyclin D immunoprecipitated from wheat bran-SCFA treated cells (Fig. 9C). However, p27 is a more effective inhibitor of cyclin E/cdk2 than cyclin D/cdk4, although p27 binds efficiently to both kinases,[23] so the associated p21 probably causes most of the cdk4 kinase inhibition. Similar levels of cyclin D1 (Fig. 9D) were found associated with cdk4 in wheat bran fiber SCFA treated and untreated cultures, indicating that the decreased cdk4 kinase activity was not due to loss of cyclin D1 from the complexes. We infer from these experiments that SCFAs mimicking digested oat bran and pectin were less effective in inducing growth arrest than those mimicking wheat bran because digested oat bran and pectin were less effective in inducing cdk2 and cdk4 inhibitors (Figs. 2–7).

2.4. Slower Induction of cdk Inhibitors by TGFβ1 than by SCFAs Indicates TGFβ1 Does Not Mediate U4 Growth Arrest by SCFAs

We next tested the hypothesis that the mechanism used by fiber SCFA mixtures for induction of the cdk inhibitors p21 and p27 was indirect, through induction of TGFβ1 which then activated transcription of both genes. Initially we determined that 4–6 fold increases in TGFβ1 protein levels could be induced in U4 pre-enterocytic cells

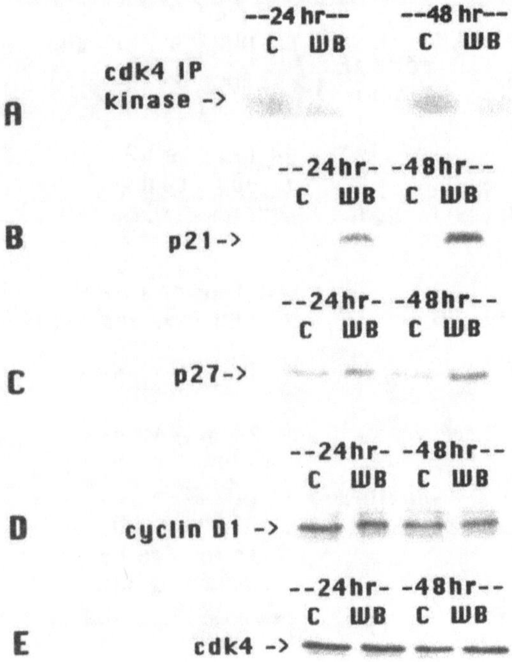

Figure 9. Analysis of cdk4/cyclin complexes immunoprecipitated from U4 cells which had been treated for 24 and 48 hours with a mixture of SCFAs mimicking digestion products of wheat bran (WB); decreased kinase activity is shown due to increased abundance of cdk inhibitors in the cdk4/cyclin D1 complex. **A**. immune complex kinase reaction using histone H1 as substrate; **B–E**, western blots of cdk4/cyclin complexes for the cdk inhibitor p21cip1 (**B**), the cdk inhibitor p27kip1 (**C**), cyclin D1 (**D**), and cdk4 (**E**), the control for equal immunoprecipitations.

by treatment for 48 hours with either 5–40 mM ammonium acetate, 1–25 mM sodium propionate, and 0.1–15 mM sodium butyrate (data not shown). Optimal concentrations of these SCFAs for 48 hours also inhibited cell growth as assayed by incorporation of a 3 hour pulse of ^{3}H-TdR or by measuring mass cell growth by incorporation of tetrazolium dye.[24] Both measurements showed that 40 mM acetate and 5 mM butyrate each inhibited growth an average of 52% while 15 mM propionate inhibited growth an average of 37%. Thus each SCFA individually elevated TGFβ1 protein levels and inhibited cell growth.

We then compared the time courses for TGFβ1 and p21 induction in U4 cultures parallel to those used to assay for p21 induction. TGFβ1 protein abundance was assayed by western blotting after 6, 24, 48, and 72 hours of treatment of U4 cells by a series of agents: SCFAs at their optimal concentration for inducing U4 cell growth arrest [40 mM acetate (A), 5 mM butyrate (B), 15 mM propionate (P)], a mixture of these three SCFA at optimal concentrations (OP), pectin SCFA mixture (PC), wheat bran SCFA mixture (W), oat bran SCFA mixture (O), and cellulose SCFA mixture (C) versus untreated control (CT) cells. Little increase in the low endogenous TGFβ1 levels was seen after 6 hr of treatment (data not shown), while acetate, butyrate, propionate, and the SCFAs mimicking digested pectin, wheat bran and oat bran induced a 5–10 fold increase in TGFβ1 abundance by 24 hours (Fig. 2). An increase in TGFβ1 abundance was observed for at least 72 hours, while control levels of TGFβ1 in U4 cells also increased due to autoinduction, as seen in many cell types.[25] If the mechanism for induction of p21cip1 by SCFAs was biphasic, with induction of TGFβ1 occurring first by SCFAs, then TGFβ1 in turn inducing transcription of p21cip1, there would be a delay in induction of p21 until TGFβ1 protein had accumulated. In contrast, if SCFAs directly induced p21cip1 transcription as well as TGFβ1 transcription, a similar time course for

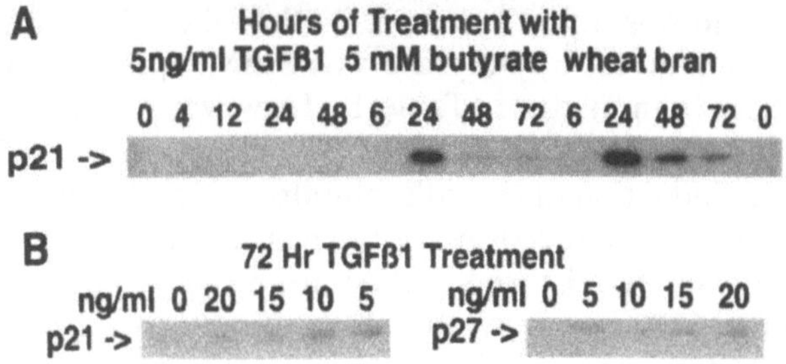

Figure 10A. Western blot showing abundance of the cdk inhibitor p21cip1 after treatment of U4 cells for 0, 6, 12, 24, and 48 hours with 5 ng/ml TGFβ1, and 6, 24, 48, and 72 hours with either 5 mM sodium butyrate or a mixture of SCFAs mimicking digestion products of wheat bran, the latter inducing a 30-fold increase in p21 levels. **B.** Western blots showing abundance of the cdk inhibitors p21cip1 and p27kip1 after treatment of U4 cells for 72 hours with 0–20 ng/ml TGFβ1, with TGFβ1 inducing only a modest 2–3 fold increase in p21 and p27 levels.

induction of p21cip1 and TGFβ1 would occur. A similar time course was observed (Fig. 2). To confirm this observation, we next compared the time courses for induction of p21 in parallel cultures by 5 ng/ml TGFβ1, 5 mM butyrate and the SCFA mixture mimicking wheat bran (Fig. 10A). No detectable induction of p21 was seen by a 4–48 hour treatment with TGFβ1. 72 hours of exposure of U4 cells to TGFβ1 was needed for a modest 2 to 3-fold induction of p21 and of p27 (Fig. 10B; p27 0–48 hr time course not shown). Butyrate and the wheat bran SCFA mixture each induced 12 to 30-fold elevated levels of p21cip1 after 24 hours of exposure in parallel cultures to those in which TGFβ1 induced no detectable p21 (Fig. 10), leading us to conclude that the wheat bran SCFA mixture does not induce cdk inhibitors through an intermediate induction of TGFβ1.

3. DISCUSSION AND CONCLUSIONS

Considerable epidemiological and experimental evidence has accumulated over the past several years indicating that intake of foods high in fiber might decrease the risk of colon cancer.[26] A recent clinical study demonstrated that dietary wheat bran fiber and supplemental calcium were associated with statistically significant decreases in fecal bile acid concentrations, a measure of colon cancer risk, in patients with a history of colon adenoma resection.[27] The mechanism of the protective effect of wheat bran fiber is not known, but may be due to a combination of factors, including the capacity of undigested fiber to bind bile acids and to decrease transit time, and the ability of SCFAs released from digested fiber to inhibit colonocyte growth.[12]

In the current study we have demonstrated a possible mechanism by which wheat bran fiber and to a lesser extent oat bran fiber and pectin fiber protect against colon cancer development. The SCFA mixtures which mimic the digestion products of wheat bran, oat bran, and pectin all induce growth arrest by induction of the cyclin dependent kinase inhibitors p21 and p27. The relative efficacies of the SCFA mixtures in inducing cdk inhibitors and growth arrest were wheat bran > oat bran > pectin > cellulose and directly parallel to the concentration of butyrate in each mixture.

However, butyrate alone was not as effective as the wheat bran mixture. The optimal concentration for growth arrest by butyrate was 5 mM, approximately the concentration found in the wheat bran mixture (Table 1). However, 5 mM butyrate for 48 hours induced less growth arrest of U4 cells than the wheat bran mixture, 52% vs 89%, probably due to the smaller induction of the cdk inhibitor p21cip1 by butyrate (Fig. 2). Since propionate and acetate also induced growth arrest and p21cip1, although to a lesser extent than butyrate (Fig. 2), all three SCFAs found in wheat bran digests are very likely to contribute to its efficacy. It is interesting to note that a synthetic combination of the three SCFAs at their optimal concentrations for inducing growth arrest, 5 mM butyrate, 15 mM propionate, and 40 mM acetate, was an average of 1.5 fold more effective than the wheat bran SCFA mixture in inducing p21cip1 (Fig. 2). Wheat bran may be the best natural fiber for inducing cdk inhibitors, but the 3.3 mM propionate and 25 mM acetate in wheat bran digests are not as high as the 15 mM propionate and 40 mM acetate in the optimal mixture.

While these SCFA mixtures which mimic the digestion products of wheat bran and other fibers also induce TGFβ1, data were shown demonstrating that TGFβ1 production does not play a major role in cdk inhibitor induction and the subsequent growth arrest of colonocytes by SCFA mixtures. TGFβ1 and butyrate activate the p21cip1 gene promoter through overlapping sites, TGFβ1 through a sequence −71 to −86 basepairs upstream of the transcription start site[28] while the butyrate-responsive elements are two SP1 sites at −82 and −69.[29] The much longer time course of induction of p21cip1 by TGFβ1 compared to butyrate suggests induction of an intermediate signaling molecule by TGFβ1, possibly insulin-like growth factor-binding protein 3, which has been shown to be necessary for inhibition of breast cancer cell growth by TGFβ.[30]

Induction of TGFβ1 by SCFAs may increase colonocyte maturation. We had found in earlier studies that TGFβ1 treatment increased cell surface abundance of integrin α-chain, a marker of enterocytic differentiation in U4 cells.[11] Integrin α-chain forms a heterodimer with integrin β1-chain, which mediates binding to collagen I. TGFβ1-treated U4 cells, as expected following the increases in integrin chain abundance, display increased binding to collagen I.[11] Collagen I binding is an initial step in differentiation of this cell type. The major role of TGFβ1 in colon enterocytic cell differentiation may be to induce cell maturation by increasing α2 integrin expression[11] and by correctly glycosylating the integrin β1-chain so it can be transported to the cell surface and form a heterodimer with the integrin α2-chain (Bellis and Friedman, manuscript in preparation).

REFERENCES

1. Vogelstein, B. and Kinzler, K. p53 function and dysfunction. *Cell* 70:523–526, 1992.
2. Markowitz, S., Wang, J., Myeroff, L., Parsons, R., Sun, L., Lutterbaugh, J., Fan, R., Zborowska, E., Kinzler, K., Vogelstein, B., Brattain, M., and Willson, J. Inactivation of the typeII TGFβ receptor in colon cancer cells with microsatellite instability. *Science* 268:1336–1338, 1995.
3. Massague, J. TGFβ signaling: receptors, transducers, and mad proteins. *Cell* 85:947–950, 1996.
4. Hahn, S., Schutte, M., Hoque, A.T.M.S., Moskaluk, C., da Costa, L., Rozenblum, E., Weinstein, C., Fischer, A., Yeo, C., Hruban, R.H., and Kern, S. DPC4, a candidate tumor suppressor gene at human chromosome 18qa21.1. *Science* 271:350–353, 1996.
5. Barnard, J., Lyons, R., and Moses, L. The cell biology of TGFβ. *Biochem. Biophys. Acta.* 1032:79–87, 1990.

6. Pelton, R., Saxena, B., Jones, M., Moses, H., and Gold, L. Immunohistochemical localization of TGFβ1, TGFβ2, and TGFβ3 in the mouse embryo: expression patterns suggest multiple roles during embryonic development. *J. Cell Biol.* 115:1091–1105, 1991.
7. Migdalska, A., Molineux, G., Demuynck, H., Evans, G., Ruscetti, F., and Dexter, T. Growth inhibitory effects of TGFβ1 *in vivo*. *Growth Factors* 4:239–245, 1991.
8. Hannon, G. and Beach, D. p15INK4B is a potential effector of TGFβ-induced cell cycle arrest. *Nature* 371:257–260, 1994.
9. Li, C.-Y., Saurdet, L., and Little, J. Potential role of waf1/cip1/p21 as a mediator of TGFβ cytoinhibitory effect, *J. Biol. Chem.* 270:4971–4974, 1995.
10. Reynisdottir, I., Polyak, K., Iavarone, A., and Massague, J. Kip/cip and ink4 cdk inhibitors cooperate to induce cell cycle arrest in response to TGFβ. *Genes Development* 9:1831–1845, 1995.
11. Hafez, M., Hsu, S., Yan, Z., Winawer, S., and Friedman, E. Two roles for TGFβ1 in colon enterocytic differentiation. *Cell Growth Differ.* 3:753–762, 1992.
12. Lupton, J. and Kurtz, P. Relationship of colonic luminal short-chain fatty acids and pH to in vivo cell proliferation in rats. *J. Nutr.* 123:1522–1530, 1993.
13. Zhang, J. and Lupton, J. Dietary fibers stimulate colonic cell proliferation by different mechanisms at different sites. *Nutr. Cancer* 22:267–276, 1994.
14. Boffa, L., Lupton, J., Marianai, M., Ceppi, M., Newmark, H., Scalmati, A., and Lipkin, M. Modulation of colonic epithelial cell proliferation, histone acetylation, and luminal short chain fatty acids by variation of dietary fiber (wheat bran) in rats. *Cancer Res.* 52:5906–5912, 1992.
15. Lee, D.-Y., Chapkin, R., and Lupton, J. Dietary fat and fiber modulate colonic cell proliferation in an interactive site-specific manner. *Nutr. Cancer* 20:107–118, 1993.
16. El-Deiry, W., Tokino, T., Velculescu, V., Levy, D., Parsons, R., Trent, J., Lin, D., Mercer, W., Kinzler, K., and Vogelstein, B. Waf1, a potential mediator of p53 tumor suppression. *Cell* 75:817–825, 1993.
17. Harper, J., Adami, G., Wei, N., Keyomarsi, K., and Elledge, S. The p21 cdk-interacting protein cip1 is a potent inhibitor of G1 cyclin-dependent kinases. *Cell* 75:805–816, 1993.
18. Lee, M.H., Reynisdottir, I., and Massague, J. Cloning of p57kip2 a cyclin dependent kinase inhibitor with unique domain structure and tissue distribution. *Genes Development* 9:639–649, 1995.
19. Hirai, H., Roussel, M.F., Kato, J.-Y., Ashmun, R.A., and Scherr, C.J. ink4 proteins p18 and p19 are specific inhibitors of the cyclin D-dependent kinases cdk4 and cdk6. *Mol. Cell Biol.* 15:2672–2681, 1995.
20. Orlow, I., Iavarone, A., Crider-Miller, S., Bonilla, F., Latres, E., Lee, M., Gerald, W., Massague, J., Weissman, B., and Cordon-Cardo, C. Cyclin-dependent kinase inhibitor p27kip2 in soft tissue sarcomas and Wilms' tumors. *Cancer Res.* 56:1219–1221, 1996.
21. Nigg, E. Cyclin-dependent kinase 7: at the cross-roads of transcription, DNA repair, and cell cycle control? *Curr. Opin. Cell Biol.* 8:312–317, 1996.
22. Higashi, H., Suzuki-Takahashi, I., Saitoh, S., Segawa, K., Taya, Y., Okuyama, A., Nishimura, S., and Kitawgawa, M. Cdk2 forms an inactive complex with cyclin D1 since cdk2 associated with cyclin D1 is not phosphorylated by cdk2-cyclin H. *Eur. J. Biochem.* 237:460–7, 1996.
23. Polyak, K., Lee, M.-H., Erdjument-Bromage, H., Koff, A., Roberts, J., Tempst, P., and Massague, J. Cloning of p27kip1, a cyclin-dependent kinase inhibitor and a potential mediator of extracellular antimitogenic signals. *Cell* 78:59–66, 1994.
24. Schroy, P., Rifkin, J., Coffey, R., Winawer, S., and Friedman, E. Role of TGFβ1 in induction of colon carcinoma differentiation by HMBA. *Cancer Res.* 50:261–265, 1990.
25. Van Obberghen-Schilling, E., Rochen, N.S., Flanders, K.C., Sporn, M., and Roberts, A. TGFβ1 positively regulates its own expression in normal and transformed cells. *J. Biol. Chem.* 263:7741–7746, 1988.
26. Nixon, D. Diet and chemoprevention of colon polyps and colorectal cancer. *Sem. Surg. Oncol.* 11:411–415, 1995.
27. Alberts, D., Titenbaugh, C., Story, J., Aickin, M., Rees-McGee, S., Buller, M., Atwood, J., Phelps, J., Ramanujam, P., Bellapravalu, S., Patel, J., Bexttinger, L., and Clark, L. Randomized, double-blinded, placebo-controlled study of effect of wheat bran fiber and calcium on fecal bile acids in patients with resected adenomatous colon polyps. *J. Natl. Cancer Inst.* 88:81–92, 1996.
28. Datto, M., Li, Y., Panus, J., Howe, D., Xiong, Y., and Wang, X. TGFβ induces the cdk inhibitor p21 through a p53-independent mechanism. *Proc. Natl. Acad. Sci. USA* 92:5545–5549, 1995.
29. Nakano, K., Mizuno, T., Sowa, Y., Orita, T., Yoshino, T., Okuyama, Y., Fujita, T., Ohtani-Fujita, N., Matsukawa, Y., Tokino, T., Yamagishi, H., Oka, T., Nomura, H., and Sakai, T. Butyrate activates the

*waf*1/*cip*1 gene promoter through SP1 sites in a p53 negative human colon cancer cell line. *J. Biol. Chem.* 272:22199–22206, 1997.
30. Oh, Y., Muller, H., Ng, L., and Rosenfeld, R. TGFβ induced cell growth inhibition in human breast cancer cells is mediated through insulin-like growth factor-binding protein 3 action. *J. Biol. Chem.* 270:13589–13592, 1995.

8

MECHANISMS BY WHICH ENERGY RESTRICTION INHIBITS CARCINOGENESIS

Henry J. Thompson, Weiqin Jiang, and Zongjian Zhu

Center for Nutrition in the Prevention of Disease
AMC Cancer Research Center
1600 Pierce Street
Lakewood, Colorado 80214

ABSTRACT

Cancer that occurs at numerous organ sites, including the colon and breast, is inhibited by energy restriction, and the inhibition is proportional to the degree of restriction imposed. In an effort to identify the mechanism(s) by which energy restriction exerts this effect, a short term model system of experimentally induced mammary carcinogenesis was used. Given that carcinogenesis is known to involve a dysregulation of tissue size homeostasis in which cell proliferation and cell death are in dysequilibrium, we hypothesized that energy restriction exerts its effect by altering one or more aspects of cell cycle regulation. It was observed that energy restriction inhibited cell proliferation and increased cell death due to apoptosis. Thus attention was next focused on aspects of cell cycle regulation that might be affected by energy restriction. It was observed that the amount of p27 protein, one member of the Cip/Kip family of genes that are involved in cell cycle arrest, was increased dose dependently by energy restriction. Based on this and related observations, the hypothesis is advanced that energy restriction inhibits carcinogenesis, at least in part, by delaying cell cycle progression via shifting cell populations into a G(0)/G(1)state. Ongoing work indicates that corticosteroids, which are produced in increased amounts in response to energy restriction, may be involved in mediating this effect.

1. OVERVIEW

Decades of investigation have clearly established that dietary restriction, in all animal species tested thus far, increases life span and reduces the occurrence of many age-associated degenerative diseases including cancer.[1–4] Recent studies have convinc-

Colon Cancer Prevention, edited under the auspices of AICR
Kluwer Academic / Plenum Publishers, New York, 1999.

ingly demonstrated that the inhibitory effect of dietary restriction on carcinogenesis is due specifically to a reduction in the intake of calories, and that inhibition of carcinogenesis is not due to a reduction in the intake of other nutrients.[5–7] Moreover, in several studies protection against carcinogenesis by energy restriction has been shown to be directly proportional to the degree of restriction imposed.[8–9] An accumulating amount of evidence points to a specific effect of energy restriction on various growth factors, oncogenes, and tumor suppresser genes that are involved in the carcinogenic process.[10–14] Moreover, data recently reported by our laboratory fails to support that energy restriction per se is sufficient for protection against carcinogenesis to be manifest.[15] Specifically, it was observed that energy restriction plus exercise failed to inhibit carcinogenesis, despite the fact that body weight gain, carcass energy, and carcass fat of exercised and energy restricted animals were reduced to a greater extent than by energy restriction or exercise alone. These unexpected findings underscore the importance of understanding the basis for the cancer inhibitory activity of agents, such as energy restriction or exercise, that modulate energy metabolism.

2. INHIBITION OF CARCINOGENESIS BY ENERGY RESTRICTION

The experiments from our laboratory that are reviewed in this chapter were conducted using a short term model for mammary carcinogenesis.[16] This model offers significant advantages compared to other related models. Morphologically identifiable intermediate stages in the disease process comparable to those that occur in the human disease can be studied, and the disease process is compressed into a 5-week versus a 6-month time period. We judge that the results obtained using this model and that are reviewed in this paper are applicable to other cancer types, particularly cancer of the colon. Using this system in combination with a meal feeding protocol to avoid potential confounding of fasting-refeeding that is frequently introduced into energy restriction studies, we observed a reduction in the carcinogenic response that was proportional to the level of energy restriction imposed.[17] These results are summarized in composite form in Table 1. The magnitude of protection against cancer that was

Table 1. Effect of energy restriction on the carcinogenic response in the mammary gland

Energy Restriction % Control[1]	Carcinogenic Response % Control[2]
90	62.7
80	48.0
60	17.3

[1]All rats were permitted access to two meals per day. Each meal was three hours in duration. Control animals had access to an unlimited amount of diet during each meal. Restricted-fed rats had access to 90, 80, or 60% of the amount consumed by the control group each meal.

[2]These data are based on the average number of adenocarcinomas per animal induced in each group as reported in reference 17. The average number of carcinomas in each energy restricted group is expressed as the percent of the response observed in the control group. The effect of energy restriction on the occurrence of carcinomas was statistically significant ($p < 0.01$).

observed is consistent with the levels of inhibition of carcinogenesis that have been reported by others.[7,8]

3. CHEMICAL BASIS FOR PROTECTIVE ACTIVITY

An objective of the experiment reported in Table 1 was to assess the potential merit of an old hypothesis, namely that a change in adrenal function, more specifically the increased production of cortical steroids, is associated with the cancer inhibitory activity of energy restriction.[18,19] Due to the time averaged, non-invasive nature of urine collection, cortical steroid abundance in 24 hour urine collections was used to assess adrenal cortical activity. This approach avoids the well documented confounding issues associated with serum corticosteroid measurement, i.e. episodic secretion, diurnal variation, and sample collection induced stress.[20] The data reported in Table 2 show that urinary cortical steroid excretion increased in proportion to the degree of energy restriction imposed and was inversely proportional to the magnitude of the carcinogenic response that was observed. In fact, urinary levels of cortical steroids explained the individual variation in the number of carcinomas per animal irrespective of the level of energy restriction to which animals were exposed.[17] This observation strongly argues that adrenal cortical steroids are either directly involved in mediating cancer inhibitory activity or that urinary cortical steroid excretion is a biomarker for other factors that account for the cancer inhibitory activity of energy restriction.

4. TISSUE SIZE REGULATION

Carcinogenesis is characterized by a failure in the regulation of tissue size homeostasis in which a clone(s) of transformed cells achieves growth-advantage due to an increased rate of cell proliferation and/or a decreased rate of cell death in comparison to neighboring populations of cells.[21,22] The net result of this process in epithelial tissues is the development of a tumor. The objective of the experiment reported in Table 3 was to examine the effects of energy restriction on tissue size regulation. The question

Table 2. Effect of energy restriction on urinary excretion of immunoreactive cortical steroids

Energy Restriction % Control[1]	Urinary Corticosterone % Control[2]
90	156
80	219
60	363

[1]All rats were permitted access to two meals per day. Each meal was three hours in duration. Control animals had access to an unlimited amount of diet during each meal. Restricted-fed rats had access to 90, 80, or 60% of the amount consumed by the control group each meal.

[2]These data are based on the average 24 hour excretion of immunoreactive cortical steroids during the final week of the five week experiment as reported in reference 17. In this table average excretion of immunoreactive cortical steroids in urine, expressed as ug/kg body weight per day, in each energy restricted group is expressed as the percent of the response observed in the control group. The effect of energy restriction on urinary excretion of immunoreactive urinary cortical steroid excretion was statistically significant ($p < 0.01$).

Table 3. Effect of energy restriction on body weight and mammary gland and mammary carcinoma size

Energy restriction[1] % Control	Body weight[2] % Control	Mammary ductal extension[3] % Control	Mammary epithelial area[4] % Control	Mammary carcinoma volume[5] % Control
90	85.2	86.2	74.3	21.0
80	79.2	83.6	63.4	11.1
60	68.5	64.9	43.8	4.9

[1]All rats were permitted access to two meals per day. Each meal was three hours in duration. Control animals had access to an unlimited amount of diet during each meal. Restricted-fed rats had access to 90, 80, or 60% of the amount consumed by the control group each meal.
[2]These data are based on the average final body weight of each group as reported in reference 35. The average final body weight in each energy restricted group is expressed as the percent of the response observed in the control group. The effect of energy restriction on final body weight was statistically significant ($p < 0.01$).
[3]These data are based on the average distance to which mammary ductal epithelium had extended into the mammary gland fat pad in each group as reported in reference 35. The average length of extension in each energy restricted group is expressed as the percent of the response observed in the control group. The effect of energy restriction on ductal extension was statistically significant ($p < 0.01$) and was in proportion to the effects of energy restriction on final body weight.
[4]These data are based on the average amount of area occupied by mammary epithelium in each group as reported in reference 35. The average epithelial area in each energy restricted group is expressed as the percent of the response observed in the control group. The effect of energy restriction on the occurrence of carcinomas was statistically significant ($p < 0.01$). The effect of energy restriction on mammary epithelial area was greater than its effect on final body weight ($p < 0.05$).
[5]These data are based on the average volume per adenocarcinoma in each group as reported in reference 35. The average volume of carcinomas in each energy restricted group is expressed as the percent of the response observed in the control group. The effect of energy restriction on carcinoma volume was statistically significant ($p < 0.001$). The effect of energy restriction on carcinoma volume was greater than its effect on body weight ($p < 0.01$).

addressed was whether the growth of the mammary gland or mammary carcinomas was disproportionately affected by energy restriction in comparison to the effects of energy restriction on overall growth measured as body weight. The data shown in Table 3 demonstrate that overall size of an animal, measured as body weight, was reduced in a manner directly proportional to the degree of energy restriction imposed. This observation is consistent with a large body of literature.[23] Ductal extension, i.e. the linear growth of mammary gland ducts was reduced in direct proportion to the effect of energy restriction on body weight. However, the development of ductal branches as well as the growth of carcinomas was reduced to a greater extent than was the effect of energy restriction on overall growth measured as body weight. These findings imply that factors other than estrogen are likely to be involved in mediating the activity of energy restriction since ductal extension within the mammary gland is under control of estrogen. In addition these findings identify a previously unrecognized differential effect of energy restriction on mammary epithelial cells depending on the developmental processes with which they are involved. The investigation of this phenomenon may provide further clues about the mechanism(s) that account for the cancer inhibitory effects of energy restriction.

5. CELL PROLIFERATION AND APOPTOSIS

In order to further pursue the observations reported in Table 3 we extended our investigation to the effects of energy restriction on rates of cell proliferation and apoptosis in mammary epithelium in uninvolved ducts and pre-malignant and malignant mammary gland lesions. These data are summarized Table 4. Energy restriction inhib-

Table 4. Effect of energy restriction on cell proliferation and apoptosis in uninvolved mammary gland duct and in pre-malignant and malignant mammary gland lesions

	Uninvolved mammary ducts[2]		Mammary intraductal proliferations[3]		Mammary adenocarcinomas[4]	
Energy restriction[1] % Control	P	A	P	A	P	A
	% Control		% Control		% Control	
90	88	100	77.2	118	86	104
80	92	111	75.6	138	77	106
60	71	160	68.5	175	72	121

[1]All rats were permitted access to two meals per day. Each meal was three hours in duration. Control animals had access to an unlimited amount of diet during each meal. Restricted-fed rats had access to 90, 80, or 60% of the amount consumed by the control group each meal.
[2]These data are based on the average rates of cell proliferation (P) and cell death due to apoptosis (A) in mammary ductal epithelial cells not involved in morphological identifiable neoplastic foci observed in each group as reported in reference 35. The average rate of P or A in each energy restricted group is expressed as the percent of the response observed in the control group. The effects of energy restriction on the rates of cell proliferation or apoptosis were not statistically significant.
[3]These data are based on the average rates of cell proliferation (P) and cell death due to apoptosis (A) in mammary intraductal proliferations, a pre-malignant neoplasm, observed in each group as reported in reference 35. The average rate of P or A in each energy restricted group is expressed as the percent of the response observed in the control group. The effects of energy restriction on the rates of cell proliferation or apoptosis were statistically significant, $p < 0.01$ and $p < 0.04$, respectively.
[4]These data are based on the average rates of cell proliferation (P) and cell death due to apoptosis (A) in mammary adenocarcinomas observed in each group as reported in reference 35. The average rate of P or A in each energy restricted group is expressed as the percent of the response observed in the control group. The effect of energy restriction on the rates of cell proliferation was statistically signficant ($p < 0.001$) whereas the effect on apoptosis was not.

ited cell proliferation and induced apoptosis to varying degrees depending on the histology of the mammary structure studied. These data are consistent with several reports about the effects of energy restriction on cell proliferation and apoptotic cell death in other model systems.[24-27] They are also consistent with "dual signal" hypothesis of cell number homeostasis; this hypothesis proposes that the pathways of proliferation and apoptosis are coupled, and implies that a proliferating somatic cell survives only if it receives appropriate survival signals.[28] According to this hypothesis, proliferating cells in somatic tissues exist on a "knife-edge" between survival and death, with survival cytokines constituting a homeostatic mechanism to control inappropriate cell expansion. Our findings of increased apoptosis and cell proliferation in adenocarcinomas of control animals, which energy restriction further increased or decreased, respectively, is consistent with this notion.

6. EXPRESSION OF CYCLIN D1 AND p27

Because of these observations we searched for potential interrelationships among the regulation of the cell cycle and adrenal cortical steroids. Based on reports that cortical steroids can modulate the expression of both cyclin D1 and p27,[29,30] we investigated the effects of energy restriction on the levels of these two proteins within cells involved in mammary carcinogenesis. As shown in Table 5, cyclin D1 was decreased and p27 was increased dose-dependently by energy restriction in terms of the number of cells expressing these proteins. Clearly the magnitude of the effect of energy restriction on p27 protein expression greatly exceeded the effect on cyclin D1. The finding

Table 5. Effect of energy restriction on the percent of cells staining postive for the expression of cyclin D1 and p27

	Uninvolved mammary ducts[2]		Mammary intraductal proliferations[3]		Mammary adenocarcinomas[4]	
Energy restriction[1] % Control	D1	p27	D1	p27	D1	p27
	% Control		% Control		% Control	
90	58.9	303	86.3	285	85.5	700
80	44.8	584	78.1	451	59.1	971
60	41.5	601	75.3	619	56.9	1295

[1]All rats were permitted access to two meals per day. Each meal was three hours in duration. Control animals had access to an unlimited amount of diet during each meal. Restricted-fed rats had access to 90, 80, or 60% of the amount consumed by the control group each meal.
[2]These data are based on the percent of cells positive for cyclin D1 (D1) and cyclin dependent kinase inhibitor p27 (p27) in mammary ductal epithelial cells not involved in morphological identifiable neoplastic foci observed in each group as reported in reference 36. The average percent of D1 or p27 positively labeling cells in each energy restricted group is expressed as the percent of the response observed in the control group. The effect of energy restriction on the number of cyclin D1 positive cells was not statistically significant whereas the effect of energy restriction on the number of p27 positive staining cells was highly significant, $p < 0.001$.
[3]These data are based on the percent of cells positive for cyclin D1 (D1) and cyclin dependent kinase inhibitor p27 (p27) in mammary intraductal proliferations, a pre-malignant mammary neoplasm, observed in each group as reported in reference 36. The average percent of D1 or p27 positively labeling cells in each energy restricted group is expressed as the percent of the response observed in the control group. The effect of energy restriction on the number of cyclin D1 positive cells was not statistically significant whereas the effect of energy restriction on the number of p27 positive staining cells was highly significant, $p < 0.003$.
[4]These data are based on the percent of cells positive for cyclin D1 (D1) and cyclin dependent kinase inhibitor p27 (p27) in mammary adenocarcinomas observed in each group as reported in reference 36. The average percent of D1 or p27 positively labeling cells in each energy restricted group is expressed as the percent of the response observed in the control group. The effect of energy restriction on the number of cyclin D1 positive cells was statistically significant ($p < 0.004$) as was the effect of energy restriction on the number of p27 positive staining cells ($p < 0.001$).

that energy restriction has a dominant effect on p27 protein expression patterns carries with it the potential of identifying a cause and effect relationship. Among the intriguing aspects of this finding is that loss of p27 protein expression in p27 knock out mice results in organomegaly,[31] a finding that would be predicted if p27 played an important role in maintenance of tissue size.[21,22] It is also of interest that p27 is an inhibitor of the cyclin D1/cdk 4 and cyclin E/cdk 2 complexes,[32,33] for which there is mounting evidence of dysregulation in mammary carcinogenesis.[34]

7. SUMMARY

Energy restriction is one of the most potent physiological inhibitors of the carcinogenic process identified to date, and energy restriction inhibits carcinogenesis in many epithelial target organs. This situation suggests that a common mechanism could, at least in part, account for energy restriction mediated protection in different tissues. While it is probable that multiple mechanisms are involved, we interpret the data presented in this chapter to indicate that energy restriction may work in part by delaying the entry of carcinogen initiated cells as well as their none transformed counterparts into the cell cycle. The induction of p27 could be an important molecular effector of this activity. This hypothesis should provide the basis for designing new experiments that ultimately will elucidate the mechanistic basis for the protective effects of energy restriction in preventing the development of cancer.

ACKNOWLEDGMENT

This work was supported by PHS grant CA 52626 from the National Cancer Institute.

REFERENCES

1. Masoro, E.J. Dietary restriction. *Exp. Gerontol* 30:291–298, 1995.
2. Kritchevsky, D. Diet in heart disease and cancer. *Adv. Exp. Med. Biol.* 369:201–209, 1995.
3. Kritchevsky, D. Caloric restriction and experimental carcinogenesis. *Adv. Exp. Med. Biol.* 322:134–141, 1992.
4. Weindruch, R. and Walford, R.L. The retardation of aging and disease by dietary restriction. Springfield, IL: Charles C Thomas, Publisher, 1988.
5. Birt, D.F., Pelling, J.C., White, L.T., Dimitroff, K., and Barnett, T. Influence of diet and calorie restriction on the initiation and promotion of skin carcinogenesis in the SENCAR mouse model. *Cancer Res.* 51:1851–1854, 1991.
6. Klurfeld, D.M., Weber, M.M., and Kretchevsky, D. Inhibition of chemically induced mammary and colon tumor promotion by calorie restriction in rats fed increased dietary fat. *Cancer Res.* 47:2759–2761, 1987.
7. Welsch, C.W. Relationship between dietary fat and experimental mammary tumorigenesis: a review and critique. *Cancer Res.* (Suppl) 52:2040s–2048s, 1982.
8. Welsch, C.W. Dietary fat, calories, and mammary gland tumorigenesis. *Adv. Exp. Med. Biol.* 322:203–222, 1992.
9. Ip, C. Quantitative assessment of fat and calorie as risk factors in mammary carcinogenesis in an experimental model. *Prog. Clin. Biol. Res.* 346:107–117, 1990.
10. Husting, S.D., Switzer, B.R., French, J.E., and Kari, F.W. The growth hormone: insulin-like growth factor 1 axis is a mediator of diet restriction-induced inhibition of mononuclear cell leukemia in Fischer rats. *Cancer Res.* 53:2750–2757, 1993.
11. Ruggeri, B.A., Klurfield, D.M., Kritchevsky, D., and Furlanetto, R.W. Calorie restriction and 7, 12-dimethylbenz(α)anthracene-induced mammary tumor growth in rats: alterations in circulating insulin, insulin-like growth factors 1 and 2, and epidermal growth factor. *Cancer Res.* 49:4130–4134, 1989.
12. Boissonneault, G.A. Calories and carcinogenesis: modulation by growth factors. In: I. Rowland (ed.), Nutrition, Toxicity, and Cancer, pp. 413–437. Boca Raton, FL: CRC Press, 1991.
13. Fernandes, G., Chandrasekar, B., Troyer, D.A., Venkatraman, J.T., and Good, R.A. Dietary lipids and calorie restriction affect mammary tumor incidence and gene expression in mouse mammary tumor virus/v-Ha-ras transgenic mice. *Proc. Natl. Acad. Sci. USA* 92:6494–6498, 1995.
14. Hursting, S.D., Perkins, S.N., and Phang, J.M. Calorie restriction delays spontaneous tumorigenesis in p53-knockout transgenic mice. *Proc. Natl. Acad. Sci. USA* 91:7036–7040, 1994.
15. Gillette, C.A., Zhu, Z., and Thompson, H.J. Energy availability and mammary cancer; a comparison of the effects of calorie restriction and exercise. *Carcinogenesis* (In preparation).
16. Thompson, H.J., McGinley, J.N., Rothhammer, K., and Singh, M. Rapid induction of mammary intraductal proliferations, ductal carcinoma in situ and carcinomas by the injection of sexually immature female rats with 1-methyl-1-nitrosourea. *Carcinogenesis* 16:2407–2411, 1995.
17. Zhu, Z., Haegele, A., and Thompson, H.J. Effect of caloric restriction on premalignant and malignant stages of mammary carcinogenesis. *Carcinogenesis* 18:1007–1012, 1997.
18. Boutwell, R.K., Brush, M.K., and Rusch, H.P. Some physiological effects associated with chronic caloric restriction. *Am. J. Physiol.* 154:517–524, 1948.
19. Pashko, L.L. and Schwartz, A.G. Reversal of food restriction-induced inhibition of mouse skin tumor promotion by adrenalectomy. *Carcinogenesis* 10:1925–1928, 1992.
20. Zadik, Z., De Lacerda, L., De Carmargo, L.A., Hamilton, B.P., Migeon, C.J., and Kowarski, A.A. A comparative study of urinary 17-hydroxycorticosteroids, urinary free cortisol, and the integrated concentration of plasma cortisol. *J. Clin. Endocrinol. Metab.* 51:1099–1101, 1980.
21. Thompson, H.J., Strange, R., and Schedin, P.J. Apoptosis in the genesis and prevention of cancer. *Cancer Epidemiol. Biomarkers Prev.* 1:597–602, 1992.
22. Steel, G.G. Growth rate of tumors. In: Steel, G.G., (ed.). Growth Kinetics of Tumors. Oxford: Clarendon, 1977, 6.

23. Freedman, L.S., Clifford, C., and Messina, M. Analysis of dietary fat, calories, body weight, and the development of mammary tumors in rats and mice: a review. *Cancer Res.* 50:5710–5719, 1990.
24. Engelman, R.W., Day, N.K., and Good, R.A. Calories, parity, and prolactin influence mammary epithelial kinetics and differentiation and alter mouse mammary tumor risk. *Cancer Res.* 53:1188–1194, 1993.
25. Steinbach, G., Kumar, S.P., Reddy, B.S., Lipkin, M., and Holt, P.R. Effects of caloric restriction and dietary fat on epithelial cell proliferation in rat colon. *Cancer Res.* 53:2745–2749, 1993.
26. Grasl-Kraupp, B., Bursch, W., Ruttkay-Nedecky, B., Wagner, A., Lauer, B., and Schulte-Hermann, R. Food restriction elimicates preneoplastic cells through apoptosis and antagonizes carcinogenesis in rat liver. *Proc. Natl. Acad. Sci. USA* 91:9995–9999, 1994.
27. James, S.J. and Muskhelishvili, L. Rates of apoptosis and proliferation vary with caloric intake and may influence incidence of spontaneous hepatoma in C57BL/6 × C3H F1 mice. *Cancer Res.* 54:5508–5510, 1994.
28. Harrington, E.A., Bennett, M.R., Fanidi, A., and Evan, G.I. c-Myc-induced apoptosis in fibroblasts is inhibited by specific cytokines. *EMBO J.* 13:3286–3295, 1994.
29. Hengst, L. and Reed, S.I. Translational control of p27Kip1 accumulation during the cell cycle. *Science* 271:1861–1864, 1996.
30. Nakayama, K., Ishida, N., Shirane, M., Inomata, A., Inoue, T., Shishido, N., Horii, I., Loh, D.Y., and Nakayama, K. Mice lacking p27(Kip1) display increased body size, multiple organ hyperplasia, retinal dysplasia, and pituitary tumors. *Cell* 85:707–720, 1996.
31. Kiyokawa, H., Kineman, R.D., Manova-Todorova, K.O., Soares, V.C., Hoffman, E.S., Ono, M., Khanam, D., Hayday, A.C., Frohman, L.A., and Koff, A. Enhanced growth of mice lacking the cyclin-dependent kinase inhibitor function of p27(Kip1). *Cell* 85:721–732, 1996.
32. Polyak, K., Kato, J.Y., Solomon, M.J., Sherr, C.J., Massague, J., Roberts, J.M., and Koff, A. $p27^{Kip1}$, a cyclin-Cdk inhibitor, links transforming growth factor-beta and contact inhibition to cell cycle arrest. *Genes Dev.* 8:9–22, 1994.
33. Sherr, C.J. and Roberts, J.M. Inhibitors of mammalian G1 cyclin-dependent kinases. *Genes Dev.* 9:1149–1163, 1995.
34. Fredersdorf, S., Burns, J., Milne, A.M., Packham, G., Fallis, L., Gillett, C.E., Royds, J.A., Peston, D., Hall, P.A., Hanby, A.M., Barnes, D.M., Shousha, S., O'Hare, M.J., and Lu, X. High level expression of p27(kip1) and cyclin D1 in some human breast cancer cells: inverse correlation between the expression of p27(kip1) and degree of malignancy in human breast and colorectal cancers. *Proc. Natl. Acad. Sci. USA* 94:6380–6385, 1997.
35. Zhu, Z., Jiang, W., and Thompson, H.J. Effect of energy restriction on tissue size regulation during chemically induced mammary carcinogenesis. *Carcinogenesis*, (In press) 1998.
36. Zhu, Z., Jiang, W., and Thompson, H.J. Effect of energy restriction on the expression of cyclin D1 and p27 during pre-malignant and malignant stages of chemically induced mammary carcinogenesis. *Molec. Carcinogenesis* (In press) 1998.

DIETARY INTERVENTION STUDIES OF COLORECTAL CANCER

Arthur Schatzkin

Division of Cancer Etiology and Genetics
National Cancer Institute
9000 Rockville Pike, EPN 211
Bethesda, Maryland 20892

This presentation will address three questions: 1) Why is dietary modification a viable strategy for preventing colorectal cancer? 2) Why are intervention studies an important component of cancer prevention research? 3) What is the current 'state of the evidence' from dietary intervention studies of colorectal cancer?

DIETARY MODIFICATION TO PREVENT COLORECTAL CANCER

Evidence from many studies in several disciplines suggests that dietary factors play an important role in colorectal carcinogenesis. Experiments in rodents administered potent carcinogens show that modulation of dietary factors can alter colon tumor burden.[1] In human metabolic studies, changing dietary fat intake can influence the production of secondary bile acids or the proliferative activity of colorectal epithlium.[2] Diet is a reasonable explanation for ecologic data: colorectal cancer rates vary widely internationally and diet varies markedly from country to country; disease rates have increased dramatically in Japan and China paralleling major dietary shifts in those countries; incidence rates among migrants tend to converge from those of the country of origin to those of the country of destination, consistent with the marked dietary changes accompanying acculturation.[3] Numerous observational epidemiologic studies implicate dietary factors in the genesis of colorectal malignancy.[4] Finally, food and its metabolites come into direct contact with gastrointestinal mucosa and molecular elements of gastrointestinal metabolism may influence cell kinetics or alter genomic integrity.

Colon Cancer Prevention, edited under the auspices of AICR
Kluwer Academic / Plenum Publishers, New York, 1999.

Table 1. Dietary factors linked to colorectal cancer

Dietary factors throught to increase risk:
fat
red meat
browned meat (cooked at high temperature—with a high concentration of heterocyclic amines)
Dietary factors thought to decrease risk:
vitamins (A, carotenoids, E, folate)
minerals (calcium, selenium)
dietary fiber
vegetables
multifactorial eating plan (e.g., one low in fat and red meat, high in fiber and vegetables)

The diet-colorectal cancer literature suggests that several specific dietary factors, each amenable to study within a randomized trial context, may influence colorectal carcinogenesis (see Table 1):

THE VALUE OF INTERVENTION STUDIES

In spite of the abundance of evidence from observational epidemiology and animal and human metabolic studies, considerable uncertainty persists about whether a given dietary change will truly reduce the incidence of colorectal cancer. This is where intervention studies (randomized trials) can be of particular value.

Trials have several strengths. They can directly test change. They allow the investigator to control exposure, and, in particular, to artificially create a wide "range" of exposure. And, what is especially important, the randomized design minimizes confounding that potentially plagues observational studies.

Trials do have their limitations, though. Once the investigator decides on the intervention (treatment), this generally remains fixed for the duration of the trial, even if new information should arise during the course of the trial suggesting that the intervention was not optimal (in terms of dose for supplements or type of dietary change). The duration of exposure (that is intervention) in a trial is relatively short, usually no more than 5 years or so in most intervention studies (the Women's Health Initiative is an exception—see below). For trial participants in the intervention arm, adherence can be problematic; in the control arm, "drop-in" or "contamination" can become troublesome. Both lack of adherence and contamination compromise the power of a trial. Trials, of course, can be quite expensive. Moreover, especially when dealing with rather modest treatment effects, results from multiple trials may be inconsistent.

Finally, there is an inferential asymmetry associated with trials. Results from a positive study are especially convincing, in that the only plausible interpretations are that the intervention works (at least as well as the observed treatment effect) or the positive results were due to chance. Results from a null study, however, tend to be much less informative, given the rather large number of potential explanations: the treatment doesn't work; the intervention was less successful than it appeared (participants didn't take their pills or truly make the reported dietary changes); the intervention wasn't specified correctly (it should have been red meat rather than fat reduction, for example); the duration of the intervention was inadequate; follow-up was not long enough (especially to capture intervention effects on early events in carcinogenesis);

the intervention should have been made at an earlier time of life; the results were due to chance.

In summary, trials 1) may provide compelling evidence that dietary modulation can prevent colorectal cancer; 2) deserve to be an important component of an overall research program in colorectal cancer prevention; 3) are not a panacea.

APPROPRIATE ENDPOINTS FOR COLORECTAL CANCER PREVENTION TRIALS

Colorectal cancer, though the second leading cause of fatal malignant disease in the United States, still occurs relatively rarely. The annual incidence rate in this country for men and women combined is approximately 50/100,000, that is, about 0.05%. Therefore, intervention studies with invasive colorectal cancer end points would have to be large, long, and costly. It is not surprising, therefore, that investigators have been extremely interested in discovering surrogate endpoints for invasive cancer. Studies with surrogate end points could be smaller, faster, and cheaper than investigations with cancer as the primary outcome.

The use of noncancer end points, however, can be costly in terms of inferential strength. To see that, we need to define a surrogate end point marker, as follows: a surrogate for colorectal cancer yields a valid test of the null hypothesis of no association between intervention and colorectal cancer. That is, a study using the surrogate end point marker gives the right answer about cancer (which is, after all, what we mainly care about).

A number of cellular and biochemical markers have been proposed as surrogates for colorectal cancer.[5] The validity of such markers, though, is uncertain, as we can see by considering rectal mucosal proliferation, as measured by proliferating cell nuclear antigen (PCNA), as a potential surrogate end point marker. ("Hyperproliferation" can be reflected in an increased labeling index, the proportion of all crypt cells that are labeled, or in an upward distribution of labeled cells from the base to crypt base to luminal surface.) We assume (and this is by no means certain) that hyperproliferation is a step on the pathway to cancer. We suppose further, though, that there are two separate pathways to colorectal cancer, one through mucosal hyperproliferation, the other through additional factors like diminished apoptosis, reduced cellular adhesion, or whatever. The problem is that this alternative pathway may not be minor. To the extent that colorectal cancer does develop through this alternative route, we cannot be sure that the relation of the intervention to these alternative markers doesn't offset its effect on the surrogate. Thus, we are rather in the dark as to whether this proliferation marker is giving us the right answer about colorectal cancer. The irony of all this is that to get the evidence that a proliferation index (including a new marker that combines information on proliferation *and* apoptosis) is a valid surrogate for colorectal cancer, we'd have to integrate this marker in studies with cancer end points. And those are precisely the studies we were trying to avoid in the first place. For those interested in this general problem of surrogate end points, I strongly recommend a recent article by Fleming and DeMets,[6] in which they show that a number of biologically plausible surrogate markers do not give the right answer with regard to whether a treatment affects primary chronic disease end points in several clinical trials.

We need to consider the colorectal adenoma as a surrogate for invasive cancer, given that adenoma recurrence has become a common outcome in dietary intervention trials. The main rationale for using adenomas as a colorectal cancer surrogate is the so-called adenoma-carcinoma sequence: most cancers come from adenomas, that is, the adenoma is by-and-large a necessary precursor lesion on the pathway to malignant disease.[7] In addition, because the recurrence rate for colorectal adenomas is some two orders of magnitude greater than the incidence rate for large bowel cancer, adenoma recurrence trials can be smaller and of shorter duration than studies with cancer outcomes. Moreover, because regular endoscopic surveillance follows the discovery of an adenoma, adenoma recurrence trials can be integrated with standard clinical practice.

Although the adenoma would seem to be a stronger surrogate marker than, say, hyperproliferation, due to the fact that the adenoma is considered an obligate precursor lesion for most colorectal malignancies, we may still not be able to generalize completely from adenoma findings to cancer. Here are two scenarios in which the adenoma could prove to be a less-than-perfect surrogate for cancer:

1) We can speculate that adenomas are heterogeneous, with only a small proportion, the "bad" ones, destined to go on to cancer; the overwhelming majority of adenomas are "innocent" and do not progress to malignant transformation. Suppose that exposures A and B are jointly required for the formation of bad adenomas, whereas exposures A and C are jointly needed for the genesis of the innocent lesions. Suppose we have an adenoma recurrence trial with an intervention that works only on exposure C. We would thus reduce the pool of innocent adenomas—thereby yielding a statistically significant reduction in adenoma formation in our trial—but in fact the development of bad adenomas and cancer would be unaffected. Conversely, if our intervention worked only on exposure B, we would see at most a small, nonsignificant reduction in adenomas and conclude that the intervention was ineffective—even though the intervention truly suppressed the development of bad adenomas and cancer.

2) Most recurrent adenomas are small lesions. The development of this small lesion reflects early neoplastic changes originating in normal epithelium (or unobserved microadenomas). If the intervention factor operates later in the neoplastic process, that is, during growth of a small into a large adenoma or a large adenoma into cancer, then a null result from an adenoma recurrence trial would be misleading. (A positive trial result, though, would correctly imply that cancer would be reduced, because large adenomas and cancers derive from small adenomas.)

In summary, then, results from adenoma recurrence trials constitute strong, but not absolute, evidence regarding an intervention effect on colorectal cancer.

COMPLETED OR ONGOING DIETARY INTERVENTION STUDIES

Micronutrient Supplements (Table 2)

Two small early studies of adenoma development in familial polyposis showed mixed results for vitamins. Bussey et al. demonstrated some reduction in polyp area with vitamin C in a study with 36 participants.[8] DeCosse et al., however, found that administration of vitamins C and E did not affect adenoma development in FAP patients.[9] McKeown-Eyssen et al. found that vitamins C and E did not reduce adenoma

Table 2. Dietary intervention studies of colorectal cancer: micronutrient supplements

PI/Location	Intervention	Sample size	Results
Bussey/US	C	36 (FAP pts)	↓polyp area
DeCosse/US	C and E	62	null
McKeown-Eyssen/Canada	C and E	200	null (AR)
Roncucci/Italy	A,C,E	209	↓AR (RR = ~0.2)
Greenberg/US	2 × 2: β-carotene; C and E	864	Null (AR)
MacClennan/Australia	2 × 2 × 2: β-carotene; (also fiber suppl; low fat)	424	Null (AR)
Buring/US	E (2 × 2 w/ASA)	40,000 ♀	Ongoing (CA)
Baron/US	Folate (2 × 2 w/ASA)	1120	Ongoing (AR)
Clark/US	selenium (200 μg: yeast)	1312 (w/skin ca)	↓CA (8 vs. 19 cases)
Baron/US	calcium carbonate (3 g; 1200 mg elemental ca++)	930	↓AR (1+, ↓17%; #, ↓25%)
ECP/Europe, Israel	calcium 2 g/d; (also fiber supplement)	~800	Ongoing (AR and progression)
WHI/US	calcium (1000 mg/d) + vitamin D (400 IUs/d); in 3 × 2 × 2 with hormones, diet	67,000 ♀; (45,000 in CaD)	Ongoing (CA)

AR = adenoma recurrence; CA = cancer.

recurrence in 200 Canadian sporadic adenoma-formers.[10] In an Italian study of around 200 participants, Roncucci et al. did observe a reduction in adenoma recurrence in those study participants taking vitamins A, C, and E.[11]

In the context of this rather mixed early history of vitamin-polyp trials, Greenberg, and colleagues at Dartmouth conducted a large multi-center adenoma recurrence trial (n = 864) with a 2 × 2 factorial design, the two factors being beta-carotene and vitamins C and E.[12] Randomized trial participants had to have had at least one adenoma removed at colonoscopy; colonoscopy was repeated after 1 and 4 years of follow-up. The results of this study were essentially null for all of the vitamin supplements.

The Australian Polyp Prevention Project examined the effect on adenoma recurrence of 20 mg beta-carotene/d in a trial with 2 × 2 × 2 factorial design (the other two factors being a fiber supplement and a low-fat eating plan).[13] Randomized participants having had at least one adenoma resected underwent repeat colonoscopy after two years and 4 years of follow-up. Adenoma recurrence was not reduced in participants receiving beta-carotene.

Two large ongoing studies of vitamins are worth noting. Buring and her colleagues are conducting a study of vitamin E (600 IU every other day) and aspirin in a trial with a 2 × 2 factorial design. This study involves some 40,000 women and does have the power to detect a reduction in incident colorectal cancer. Baron and colleagues are carrying out a new, large adenoma recurrence trial (n = 1120) with a 2 × 2 factorial design, one factor being aspirin, the other folate (1 mg/d).

Clark et al. conducted a skin cancer prevention trial using selenium (200 μg, as selenized yeast).[14] This trial showed that selenium supplementation had no effect on skin cancer but, in secondary analyses, the intervention was associated with a reduction in incident colorectal cancer (8 and 19 cases, respectively, in the intervention and control arms).

Baron and colleagues looked at calcium supplementation (3g calcium carbonate; 1200mg elemental calcium) in relation to adenoma recurrence in a randomized trial with 930 participants.[15] This trial was positive. The investigators observed a 17% reduction in recurrence (of any adenoma) in the intervention, compared to the control arm. They also found that the number of recurrent adenomas was reduced by 25% less in intervention compared to control participants.

The European Cancer Prevention (ECP) Organization is currently carrying out a placebo controlled multicenter, multinational adenoma recurrence trial among some 800 persons.[16] The interventions are calcium (2g/d as calcium gluconolactate) and a fiber supplement. A unique aspect of this trial is its capacity to observe the effects of the intervention agents on small polyp progression, as not all small lesions will be removed initially.

We need to include the massive NIH-sponsored Women's Health Initiative as a calcium-colorectal cancer intervention study.[17] The WHI is a randomized controlled clinical trial of about 67,000 postmenopausal women between the ages of 50 and 79. The trial has three interventions, although women can choose to be randomized into two or three of the overlapping studies. The interventions include calcium (1000mg/d) + vitamin D (400IUs/d); hormone replacement therapy; and a low-fat eating plan. The trial has approximately a 90% power to detect a 20% reduction in the incidence of colorectal cancer.

Macronutrient Supplements (Table 3)

In the study of familial polyposis patients described above, DeCosse and his coworkers also examined the effect of a dietary fiber supplement (22.5g/d) on adenoma burden. They found no overall effect.

In the Australian Polyp Prevention Project, the investigators found that participants in the fiber supplementation (wheat bran 25g/d) had a small nonsignificant increase in overall adenoma recurrence, but a slight nonsignificant decrease for large (≥1cm) adenomas.

Alberts and his colleagues are currently completing a large (n = 1425) fiber supplement adenoma recurrence trial in Arizona.[18] Intervention arm participants are

Table 3. Dietary intervention studies of colorectal cancer: macronutrient supplements

PI/Location	Intervention	Sample size	Results
DeCosse/US	fiber supplement (22.5g/d)	62 (FAP pts)	null
MacClennan/Australia	wheat bran 25g/d; in 2 × 2 × 2 w/ β-carotene; low fat	424	overall null (AR); ↓AR (1cm +) in low fat/high fiber cell
Ritenbaugh/US	wheat bran 13.5g/d	~1400	Ongoing (AR)
ECP/Europe, Israel	fiber supplement 3.8g/d; (also calcium)	~800	Ongoing (AR and progression)
CAPP/Europe	resistant starch (2 × 2 w/ASA)	~400 (FAP pts)	Ongoing

AR = adenoma recurrence.

assigned to take 13.5 g/d of wheat bran. These results should be available in the next year.

The ECP study of about 800 participants has, in addition to the calcium intervention, a fiber supplement intervention (3–8 g/d of ispaghula husk).

European investigators have initiated the CAPP (Concerted Action on Polyposis Prevention) Study among approximately 400 FAP patients.[19] The two factors in the 2 × 2 factorial design are resistant starch and aspirin.

Dietary Change (Table 4)

In another Canadian study, 201 men and women were randomized to receive counseling on a low fat (either less than 50 g per day, 20% energy), high fiber (50 g/d) diet or to follow the customary Western, high-fat, low-fiber eating pattern.[20] (The control group also received a "placebo" fiber supplement containing 3 g dietary fiber in a 50 g package.) An intention-to-treat analysis of data collected after an average of 2 years of observation showed no significant difference in adenoma recurrence between the intervention and control arms (RR = 1.2, 95% CI 0.6–2.2).

In the Australian study, participants in the dietary intervention arm were counseled to reduce intake to ≤25% calories from fat. The investigators found a slight, nonsignificant reduction in adenoma recurrence among participants in the low fat group. In an analysis restricted just to the recurrence of large (≥1 cm) adenomas, there was a marginally significant ($p = 0.05$) 70% reduction within the low-fat group (relative risk, 0.3, 95% CI, 0.1–1.0). No participants in the combined low fat-fiber supplement group developed large adenomas over 4 years of follow-up ($p \leq 0.03$).

The National Cancer Institute is currently completing the Polyp Prevention Trial (PPT), a very large (n = 2079) multicenter, randomized, controlled trial of the effect of a low-fat, high-fiber, high fruit-and-vegetable eating plan on adenoma recurrence.[21] The PPT intervention participants were counseled to change their overall dietary pattern by meeting three goals: 20% calories from fat, 18 g fiber per 1000 kcal, 5–8 servings of fruits and vegetables. Repeat colonoscopy was carried out after 1 and four years. Recruitment for the study was completed in early 1994; results should be reported in 1999.

Table 4. Dietary intervention studies of colorectal cancer: fat reduction, multifactorial

PI/Location	Intervention	Sample size	Results
McKeown-Eyssen/Canada	low fat (20% cals) and fiber supplement (50 g/d)	201	null (AR RR = 1.2, 0.6–2.2)
MacClennan/Australia	low fat (≤25% cals); in 2 × 2 × 2 w/β-carotene; wheat bran	424	overall null (AR); ↓AR (1 cm+ only) in low fat group and low fat/high fiber cell
PPT/US	20% fat calories; 18 g/1000 kcal/d; 5–8 serv fruits-veg/d	2079	Ongoing (AR)
WHI/US	↓fat (total, 20% cals; sat., 7% cals); ↑ grains (6/d); ↑fruits/veg (5/d); in 3 × 2 × 2 with CaD, hormones	67,000 ♀; 47,000 in diet component	Ongoing (CA)

The low-fat intervention in the Women's Health Initiative is really a multifactorial one: it involves not only a reduction in fat intake (to 20% of calories from total fat, 7% of calories from saturated fat) but also an increase in grain consumption (at least 6 servings/d) and fruit, and vegetable consumption (at least 5 servings/d). Thus this study will provide a direct test of the hypothesis that this multidimensional eating pattern modification reduces the incidence of invasive colorectal cancer.

SUMMARY OF EVIDENCE TO DATE FROM DIETARY INTERVENTION STUDIES

Trials thus far suggest that there is not much of a preventive story for vitamin A or beta-carotene. Similarly, the vitamin E results have not been promising, but we await results of the large study of 40,000 women with explicit cancer outcomes. A positive calcium-polyp trial is not easily dismissed, but further trials are needed for confirmation of this finding. The selenium finding may turn out to be little more than a change *post hoc* result: it needs confirmation in an additional randomized trial.

Smaller intervention studies have shown no confirmation of the dietary fiber hypothesis; we await results from larger trials. Again, with regard to dietary fat, the smaller polyp trials provide no confirmation of the hypothesis and we will have to see whether any positive findings emerge from the large trials, the PPT, and WHI. The observational evidence for protection from fruit and vegetable (mainly vegetable) intake is strong, but empirical corroboration is lacking; the PPT and WHI may be informative. Finally, no intervention studies of meat consumption have been carried out. Such a study (the intervention could involve eliminating meat or simply altering cooking practices) might now be appropriate.

Thus, aside from largely burying the carotenoid hypothesis and breathing new life into calcium, dietary intervention studies have yet to yield informative answers for most major hypotheses. We eagerly await results of a number of larger trials with neoplastic and carcinoma endpoints.

REFERENCES

1. Zhao, L.P., Kushi, L.H., Klein, R.D., and Prentice, R.L. Quantitative review of studies of dietary fat and rat colon carcinoma. *Nutr. Cancer* 15:169–177, 1991.
2. Reddy, B.S. Diet and excretion of bile acids. *Cancer Res.* 41:3766–3768, 1981.
3. Schatzkin, A., Schiffman, M., and Lanza, E. Research priorities in large bowel cancer prevention. *Semin. Oncol.* 17:425–437, 1990.
4. Potter, J.D. Nutrition and colorectal cancer. *Cancer Causes Control* 7:127–146, 1996.
5. Wargovich, M.J. Precancer markers and prediction of tumorigenesis. *In* Young, G.P., Rozen, P., and Levin, B., eds. Prevention and early detection of colorectal cancer. Philadelphia: WB Saunders Company Ltd, 1996, pp. 89–101.
6. Fleming, T.R. and DeMets, D.L. Surrogate end points in clinical trials: are we being misled? *Ann. Intern. Med.* 125:605–613, 1996.
7. Schatzkin, A., Freedman, L.S., Dawsey, S.M., and Lanza, E. Interpreting precursor studies: what polyp trials tell us about large bowel cancer. *J. Natl. Cancer Inst.* 86:1053–7, 1994.
8. Bussey, H.J., DeCosse, J.J., Deschner, E.E., et al. A randomized trial of ascorbic acid in polyposis coli. *Cancer* 50:1434–1439, 1982.
9. DeCosse, J.J., Miller, H.H., and Lesser, M.L. Effect of wheat filer and vitamins C and E on rectal polyps in patients with familial adenomatous polyposis. *J. Natl. Cancer Inst.* 81:1290–1297, 1989.

10. McKeown-Eyssen, G., Holloway, C., Jazmaji, V., Bright-See, E., Dion, P., and Bruce, W.R. A randomized trial of vitamins C and E in the prevention of recurrence of colorectal polyps. *Cancer Res.* 48: 4701–4705, 1988.
11. Roncucci, L., DiDonato, P., Carati, L., et al. Antioxidant vitamins or lactulose for the prevention of the recurrence of colorectal adenomas. *Dis. Colon Rectum* 36:227–234, 1993.
12. Greenberg, E.R., Baron, J.A., Tosteson, T.D., et al. A clinical trial of antioxidant vitamins to prevent colorectal adenoma. *New Engl. J. Med.* 331:141–147, 1994.
13. MacLennan, R., Macrae, F., Bain, C., et al. Randomized trial of intake of fat, fiber, and beta-carotene to prevent colorectal adenomas. *J. Natl. Cancer Inst.* 87:1760–1766, 1995.
14. Clark, L.C., Combs, G.F., and Turnbull, B.W. Effects of selenium supplementation for cancer prevention in patients with carcinoma of the skin: a randomized controlled trial. *JAMA* 276:1957–1963, 1996.
15. Baron, J. Personal communication 1998.
16. Faivre, J., Doyon, F., and Boutron, M.C. For the European Cancer Prevention Colon Group. The ECP calcium fiber polyp prevention study. *Eur. J. Cancer Prev.* 1(supplement 2):83–89, 1991.
17. Women's Health Initiative Study Group. Design of the Women's health Initiative Clinical Trial and Observational Study. *Control. Clin. Trials* 19:61–109, 1998.
18. Martinez, M.E., Reid, M.E., Guillen-Rodrigues, J., et al. Design and baseline characteristics of study participants in the wheat bran fiber trial. *Cancer Epidem. Biomarkers Prev.* 7:813–816, 1998.
19. Burn, J., Chapman, P., and Eastham, E. A European double blind trial of aspirin and resistant starch in colon cancer prevention. *Arch. Dis. Childhood* 71:103–107, 1994.
20. McKeown-Eyssen, G.E., Bright-See, E., Bruce, W.R., et al. A randomized trial of a low fat high fibre diet in the recurrence of colorectal polyps. *J. Clin. Epidemiol.* 47:525–536, 1994.
21. Schatzkin, A., Lanza, E., Freedman, L.S., Tangrea, J., Cooper, M.R., Iber, F., Selby, J., Caan, B., Weissfeld, J., Burt, R., Slattery, M., Kikendall, W., and Cahill, J. The Polyp Prevention Trial. I.: rationale, design, recruitment, and baseline participant characteristics. *Cancer Epidemiol. Biomarkers Prev.* 5:375–383, 1996.

10

THE GENETICS OF HEREDITARY NON-POLYPOSIS COLORECTAL CANCER AND NON-POLYPOTIC COLON CANCER

Päivi Peltomäki

Division of Human Cancer Genetics
Comprehensive Cancer Center
The Ohio State University
Columbus, Ohio

1. INTRODUCTION

Most colorectal cancers are sporadic but the lifetime risk for this cancer increases with the number of affected individuals in the pedigree. The presence of three affected relatives from two generations is compatible with autosomal dominant inheritance and fulfils the international diagnostic ("Amsterdam") criteria for hereditary non-polyposis colorectal cancer, HNPCC.[1]

The detection of patients with hereditary predisposition is important because the disease genes usually have high penetrance and the risk may not be increased for colon cancer only but for numerous extracolonic cancers as well.[2] Furthermore, a majority of colorectal cancers from HNPCC patients as well as some 15% of sporadic colorectal cancers show microsatellite instability (MSI) as a characteristic abnormality, and the prognosis may be different in patients with MSI(+) tumors as compared to those with MSI(–) ones.[3] Finally, the detection of genetic susceptibility is important because the development of cancer at high-risk individuals may be prevented by early intervention.[4]

2. METHODS

We conducted a prospective study in which MSI analyses were used to screen for hereditary susceptibility among consecutive patients with apparently sporadic colon cancer.[5] Tumor samples from 509 patients were tested and 63 (12%) displayed MSI.

Colon Cancer Prevention, edited under the auspices of AICR
Kluwer Academic / Plenum Publishers, New York, 1999.

Normal tissues from the latter patients were tested for germline mutations in the two major HNPCC-associated DNA mismatch repair genes, MSH2 and MLH1, and a mutation was found in 10 (2% of all colorectal cancer patients). A retrospective analysis of the data showed that patients with inherited susceptibility who were detected by our molecular genetic analyses all showed one or more of the clinical hallmarks of HNPCC, namely family history of cancer (90%), multiple cancers (40%), and age at onset below 50 years (70%).

According to an international database (*http://www.nfdht.nl*) more than 200 different HNPCC-associated mutations have been described and most affect the DNA mismatch repair genes MSH2 and MLH1. Apart from DNA mismatch repair genes, another type of gene, TGFβRII, was recently implicated as a susceptibility gene to HNPCC.[6] The phenotype was atypical and no MSI was observed in tumors in agreement with the fact that TGFβRII is not known to participate in the DNA mismatch repair pathway. Most MSH2 mutations are truncating. This is also the case for MLH1 mutations but here also missense mutations (that only lead to an amino acid substitution) play an important role.

How does genotype correlate with phenotype in HNPCC? Although some correlations have been proposed in the literature,[7–9] the association of particular mutations with specific phenotypes is generally poor.

Mutations in the DNA mismatch repair genes obviously make a major contribution to the fact that most people who carry germline mutations of these genes do develop cancer during their lifetime. However, little is known about mechanisms that determine why some individuals develop colon cancer while others develop endometrial cancer, for example. Even in kindreds with identical predisposing mutations there may be remarkable variation between families and between different individuals from the same family, suggesting that additional phenotype determinants are likely to exist.

Apart from correcting synthesis errors that arise during DNA replication, the DNA mismatch repair system has recently been directly implicated in the protection of cells against environmentally induced damage. The system has been shown to recognize and process at least UV photoproducts,[10] alkylating agents,[11] and aromatic amines.[12] Starting from these observations, we studied if genetic polymorphisms in three loci involved in carcinogen metabolism, namely N-acetyltransferase 1 (NAT1) and glutathione-S-transferases mu (GSTM1) and theta (GSTT1) might modify clinical characteristics of HNPCC. The gene product of NAT1 may activate aromatic amines and allele *10 is associated with the rapid acetylation phenotype that has been suggested to increase colorectal and other cancer risk.[13] Glutathione-S-transferases detoxify electrophilic compounds and null genotypes may increase colorectal and other cancer risk.[14]

For association studies it is of crucial importance that the populations studied are as homogeneous as possible. We therefore focused on two groups of HNPCC patients from Finland, each of which showed a shared predisposition (MLH1 "mutation 1" in one group and MLH1 "mutation 2" in the other) and shared ancestry.[15] Colonic location of tumor and age at onset were assessed as possible features under modification. Looking at the location of tumors from patients with mutation 1 there was a characteristic overrepresentation of proximal tumors when all 78 tumors were analyzed as a group. However, stratification according to the presence vs. absence of the NAT1 allele *10 revealed a significant difference: in the group with the allele *10 present, the number of distal tumors exceeded that of proximal ones.[16] Furthermore, the presence of NAT1*10 was associated with lower median age at onset in both the mutation 1 and mutation 2 groups, and the null alleles of GSTM1 and GSTT1 showed similar age effects in patients with mutation 1.[16] Although additional studies are necessary to investigate these associations in greater detail our results lend preliminary support that mechanisms implicated as risk factors for sporadic colon cancer may modify phenotypic manifestations of hereditary colon cancer as well.

3. CONCLUSIONS

According to the present knowledge, DNA mismatch repair genes account for 70% of HNPCC kindreds meeting the Amsterdam criteria and showing microsatellite instability in tumors.[17] In the remaining 30% the predisposing genes are yet to be identified. DNA mismatch repair gene mutations are detected in only a minority of kindreds not meeting the Amsterdam criteria and the basis of susceptibility is unknown in 70%.[18] Genes predisposing to MSI negative familial nonpolypotic colon cancer are almost completely unknown. Future studies should address the nature of genetic predisposition in these remaining important subgroups of familial nonpolypotic colon cancer.

REFERENCES

1. Vasen, H.F.A., Mecklin, J.-P., Meera Khan, P., and Lynch, H.T. The International Collaborative Group on Hereditary Non-Polyposis Colorectal Cancer (ICG-HNPCC). *Dis. Colon Rectum* 34:424–425, 1991.
2. Aarnio, M., Mecklin, J.-P., Aaltonen, L.A., Nyström-Lahti, M., and Järvinen, H.J. Lifetime risk of different cancers in hereditary nonpolyposis colorectal cancer (HNPCC) syndrome. *Int. J. Cancer* 64: 430–433, 1995.
3. Lothe, R.A., Peltomäki, P., Meling, G.I., Aaltonen, L.A., Nyström-Lahti, M., Pylkkänen, L., Heimdal, K., Andersen, T.I., Møller, P., Rognum, T.O., Fossa, S.D., Haldorsen, T., Langmark, F., Brøgger, A., de la Chapelle, A., and Børresen, A.-L. Genomic instability in colorectal cancer: relationship to clinicopathological variables and family history. *Cancer Res.* 53:5849–5882, 1993.
4. Järvinen, H.J., Mecklin, J.-P., and Sistonen, P. Screening reduces colorectal cancer rate in families with hereditary nonpolyposis colorectal cancer. *Gastroenterol.* 108:1405–1411, 1995.
5. Aaltonen, L.A., Salovaara, R., Kristo, P., Canzian, F., Hemminki, A., Peltomäki, P., Chadwick, R.B., Kääriäinen, H., Eskelinen, M., Järvinen, H., Mecklin, J.-P., and de la Chapelle, A. Incidence of hereditary nonpolyposis colorectal cancer and the feasibility of molecular screening for the disease. *N. Engl. J. Med.* 338:1481–1487, 1998.
6. Lu, S.-L., Kawabata, M., Imamura, T., Akiyama, Y., Nomizu, T., Miyazono, K., and Yuasa, Y. HNPCC associated with germline mutation in the TGF-β type II receptor gene. *Nature Genet.* 19:17–18, 1998.
7. Vasen, H.F.A., Wijnen, J.T., Menko, F.H., Kleibeuker, J.H., Taal, B.G., Griffioen, G., Nagengast, F.M., Meijers-Heijboer, E.H., Bertario, L., Varesco, L., Bisgaard, M.-L., Mohr, J., Fodde, R., and Meera Khan, P. Cancer risk in families with hereditary nonpolyposis colorectal cancer diagnosed by mutation analysis. *Gastroenterol.* 110:1020–1027, 1996.
8. Kruse, R., Lamberti, C., Wang, Y., Ruelfs, C., Bruns, A., Esche, C., Lehmann, P., Ruzicka, T., Rutten, A., Friedl, W., and Propping, P. Is the mismatch repair deficient type of Muir-Torre syndrome confined to mutations in the MSH2 gene? *Hum. Genet.* 98:747–750, 1996.
9. Jäger, A.C., Bisgaard, M.L., Myrhøj, T., Bernstein, I., Rehfeld, J., and Nielsen, F.C. Reduced frequency of extracolonic cancers in hereditary nonpolyposis colorectal cancer families with monoallelic hMLH1 expression. *Am. J. Hum. Genet.* 61:129–138, 1997.

10. Mellon, I., Rajpal, D.K., Koi, M., Boland, C.R., and Champe, G.N. Transcription-coupled repair deficiency and mutations in human mismatch repair genes. *Science* 272:557–560, 1996.
11. Yamada, M., O'Regan, E., Brown, R., and Karran, P. Selective recognition of a cis-platin-DNA adduct by human mismatch repair proteins. *Nucleic Acids Res.* 25:491–495, 1997.
12. Li, G.-M., Wang, H., and Romano, L.J. Human MutSα specifically binds to DNA containing aminofluorene and acetylaminofluorene adducts. *J. Biol. Chem.* 271:24084–24088, 1996.
13. Bell, D.A., Stephens, E.A., Castranio, T., Umbach, D.M., Watson, M., Deakin, M., Elder, J., Hendrickse, C., Duncan, H., and Strange, R.C. Polyadenylation polymorphism in the acetyl-transferase 1 gene (NAT1) increases risk of colorectal cancer. *Cancer Res.* 55:3537–3542, 1995.
14. Zhong, S., Wyllie, A.H., Barnes, D., Wolf, C.R., and Spurr, N.K. Relationship between the GSTM1 genetic polymorphism and susceptibility to bladder, breast, and colon cancer. *Carcinogenesis* 14:1821–1824, 1993.
15. Moisio, A.-L., Sistonen, P., Weissenbach, J., de la Chapelle, A., and Peltomäki, P. Age and origin of two common MLH1 mutations predisposing to hereditary colon cancer. *Am. J. Hum. Genet.* 59:1243–1251, 1996.
16. Moisio, A.-L., Sistonen, P., Mecklin, J.-P., Järvinen, H., and Peltomäki, P. Genetic polymorphisms in carcinogen metabolism and their association to hereditary nonpolyposis colon cancer. *Gastroenterol.* (in press), 1998.
17. Liu, B., Parsons, R., Papadopoulos, N., Nicolaides, N.C., Lynch, H.T., Watson, P., Jass, J.R., Dunlop, M., Wyllie, A., Peltomäki, P., de la Chapelle, A., Hamilton, S.R., Vogelstein, B., and Kinzler, K. Analysis of mismatch repair genes in hereditary non-polyposis colorectal cancer patients. *Nature Med.* 2:169–174, 1996.
18. Nyström-Lahti, M., Wu, Y., Moisio, A.-L., Hofstra, R.M., Osinga, J., Mecklin, J.-P., Järvinen, H.J., Leisti, J., Buys, H.J., de la Chapelle, A., and Peltomäki, P. DNA mismatch repair gene mutations in 55 kindreds with verified or putative hereditary non-polyposis colorectal cancer. *Hum. Mol. Genet.* 5:763–796, 1996.

11

FAMILIAL ASSOCIATION

Randall W. Burt

University of Utah
University of Utah Health Sciences Center
Salt Lake City, Utah 84132

INTRODUCTION

Colon cancer is one of the most familial malignancies in man. There are rare syndromes of colon cancer, but also frequent familial clusters in excess of that expected by chance. The genes that are mutated in the rare syndromes are now known and the cellular processes related to these genes are being elucidated. Common familial clustering is also thought to arise from inherited susceptibility, based on segregation studies, but the underlying genes have not yet been identified. This section will summarize present knowledge of both the rare and more common forms of familial colon cancer.

INHERITED SYNDROMES OF COLON CANCER

The well known syndromes of colon cancer include familial adenomatous polyposis (FAP) and hereditary nonpolyposis colorectal cancer (HNPCC).

Familial Adenomatous Polyposis (FAP)

FAP is characterized by the early age onset of hundreds to thousands of colonic adenomatous polyps and a virtual certainty of colon cancer unless the colon is removed.[1] Polyps appear at an average age of 16 years and cancer at an average age of 39 years. Other manifestations of the condition include gastric fundic gland polyps, duodenal and small bowel adenomas, certain dental abnormalities, desmoid tumors, soft tissue tumors, and osteomas. There is also an increased risk of duodenal, CNS and thyroid and adrenal cancer, as well as hepatoblastoma. The frequency of FAP is estimated to be approximately 1 in 10,000 births. It has accounted for about 0.5% of colon cancers in the past, although this figure is now thought to be much smaller, in part due to the cancer prevention efforts applied to FAP families.

Colon Cancer Prevention, edited under the auspices of AICR
Kluwer Academic / Plenum Publishers, New York, 1999.

FAP is inherited in an autosomal dominant fashion and arises from mutations of the APC gene. Adenomatous polyps arise when the normal allele of the APC gene is inactivated by mutation or loss of heterozygosity. Virtually all disease causing mutations result in truncation of resultant protein. A main function of the APC gene (reviewed in detail in other presentations of this conference) is to bind β-catenin and target it for destruction.[2] Free β-catenin appears to be involved in activation of nuclear transcription. There is some relationship between manifestations of FAP and the location mutations in the APC gene. Profuse polyposis (>5000 adenomas in the colon), for example, occurs when mutations are found in a certain region of exon 15 of the APC gene.[3] Genetic testing is now commercially available for detection of disease causing mutations in family members at risk for FAP.[4]

Hereditary Nonpolyposis Colorectal Cancer (HNPCC)

Characteristics of HNPCC include the autosomal dominant inheritance of colon cancer predisposition, young age colon cancer (average age of diagnosis, 44 years), a predominance of proximal colonic tumors and frequent synchronous and metachronous colon cancers.[5,6] Precise clinical criteria have been established that have been helpful in identifying families with HNPCC. They are referred to as the "Amsterdam Criteria." To meet these criteria in a family there must be three first-degree relatives with colon cancer spanning at least two generations, and at least one of the cases must have been diagnosed at an age less than 50 years. Affected persons usually exhibit only a few adenomatous polyps, although on average, polyps occur at a younger age, are larger and more frequently exhibit villous histology than adenomatous polyps in population controls. Uterine cancer occurs in 40% to 60% of affected women and there is also a significant excess of ovarian, small bowel, pancreatic, gastric, and urinary tract malignancies in HNPCC.[7]

HNPCC is now known to arise from mutations of one of the six identified mismatch repair (MMR) genes.[8] The protein products of these genes are normally involved in repair of DNA errors that commonly occur with transcription—called replication errors (MMR genes are discussed in more detail in other chapters of this volume). Although six MMR genes have now been identified, mutations in two of the genes account for over 95% of families with HNPCC. There is some correlation between the specific gene mutated and the extra-colonic cancers found in affected persons. For example, an excess of urinary tract, stomach, and ovarian cancers is observed only in families where the *hMSH2* gene is mutated.[7]

When families meet the Amsterdam Criteria, almost 50% will be found to have a mutation in one of the mismatch repair genes. Families with a strong family history of colon cancer, but who do not meet these strict criteria, are found to have MMR mutations only about 8% of the time.[9] An efficient strategy for detecting MMR mutations in such families, however, has recently been suggested. Tumor tissue from a person with colon cancer is first tested for micro-satellite instability (MSI). MSI is a feature of 90% of colon cancers from persons with HNPCC but of only 15% of sporadic colorectal cancers. MSI reflects the presence of numerous DNA replication errors throughout the genome, errors which accumulate when mismatch repair is not functional. If the tumor is MSI positive, then the likelihood of finding a germline MMR mutation is much higher so that mutation detection in the MMR genes can be more efficiently undertaken.[10]

COMMON FAMILIAL CLUSTERING OF COLON CANCER CASES

Three epidemiologic studies first examined the mortality of colon cancer in relatives of persons with this malignancy, compared to population controls.[11] Each found an approximate 3.5-fold increased risk of colon cancer mortality in first-degree relatives, although the study design differed substantially between the investigations. Authors of all three studies carefully excluded persons with FAP, and monogenetic inheritance patterns were not noted among colon cancer cases. Thus, the inherited syndromes appeared unlikely to account for the familial clustering observed. A number of incident studies have now also been performed. These have each examined the incidence or occurrence of colon cancer in first-degree relatives compared to controls. Excess risk of colon cancer among relatives in these studies has varied from two- to eight-fold. Several of the studies also examined the risk of colon cancer in first-degree relatives of persons with adenomatous polyps. A two- to four-fold increased risk of colon cancer was also observed in polyp relatives. It should be noted that the nature of these studies makes it probable that the index polyps were likely large and/or symptomatic.

In an attempt to define the etiology of common familial risk we studied large kindreds that exhibited excess cases of colon cancer. Segregation analysis was used to determine if inheritance patterns were present and could explain the occurrence of adenomatous polyps and cancer in the families. Kindreds with excess colon cancer cases, but no apparent inherited syndromes, were systematically selected. Family members were examined for the presence of colonic adenomatous polyps and the prevalence of polyps in family members was compared to the prevalence in spouse controls. In the first large kindred studied, adenomas were significantly more common in relatives than in spouse controls.[12] Furthermore, segregation analysis found an autosomal dominant pattern of polyp and cancer occurrence. If environmental factors alone accounted for polyps and cancer, a more random pattern would have been expected. Thirty-three additional kindreds were then studied. The findings were similar when all kindreds were analyzed together, or when they were analyzed by ascertainment group.[13] Adenomas were also significantly more common in family members compared to spouses and segregation patterns of adenomas and cancers were autosomal dominant. Analysis further suggested that up to one-third of the population studied could have inherited susceptibility to colonic adenomas and cancer, and that one-half (or more) of the adenomas observed may have arisen, at least in part, from inherited susceptibility.

Stratifying Common Familial Risk

The precise genetic etiology of common familial clustering has not yet been clarified, although clinical studies have demonstrated that the severity of familial risk can be predicted to some degree by clinical findings in a family. Age of diagnosis and number of colon cancer cases, in particular, appear to predict the severity of cancer risk in families. A number of studies have found that the risk of colon cancer is approximately two-fold increased if a first-degree relative has this malignancy.[11] The risk is approximately doubled again if two first-degree relatives are affected. A younger age of colon cancer diagnosis is also associated with an increased risk to relatives. If the diagnosis is below age 50, then the risk to immediate relatives is increased three- to four-fold.[14] Some increased risk remains even in later decades of life. A recent prospective study with a 6 to 8 year follow-up found a relative risk of 1.79 in persons who had an immediate family member with colon cancer, and a relative risk of 2.75 if two immediate family members were affected.[15] There was a graded increase to relatives with decreasing age. A diagnosis under age 45 resulted in an almost 5-fold increased risk.

There also appears to be increased risk of colon cancer when adenomatous polyps are found in relatives.[16,17] The National Polyp Study observed an almost two-fold increased risk in parents and siblings of those with adenomatous polyps.[18] If the adenoma was diagnosed at an age less than 60 years, the risk of colon cancer was increased almost three-fold. When a patient with an adenoma also had a parent with colon cancer, siblings were found to exhibit a 3.3-fold increased risk for colon cancer. Colonoscopy studies also have generally demonstrated increased risk of adenomatous polyps in first-degree relatives of persons with large bowel malignancy. A recent study found no increased risk of adenomas in immediate family members when all adenomatous polyps were considered, but a significantly increased risk (relative risk 2.6) for advanced adenomas, i.e. those larger than one centimeter or those with villous change.[19] Common familial risk for colon cancer also appears to extend, albeit to a lesser degree, to more distant relatives, to older age relatives, and even to relatives with certain other cancers such as ovarian, uterine, and those of the breast.[20]

Genetic Etiology of Common Familial Risk

Common familial risk could arise from the exposure of family members to similar environmental factors, from moderate or low-penetrant susceptibility genes or from a combination of these. The segregation studies cited previously indicate the genetic hypothesis, but it is well known that dietary and other environmental factors also associate significantly with colon cancer risk. Genes involved in carcinogen metabolism are therefore particularly attractive candidates for low penetrant cancer predisposing genes—they could combine both inherited predisposition and dietary exposures. A polymorphism of an acetylator gene, for example, that resulted in slightly higher levels of certain gut carcinogens over a lifetime might be expected to result in an increased incidence of colon cancer in persons so affected. Polymorphisms of genes involved in acetylation, N-oxidation, and P450 metabolic rates are therefore genes that could be involved in common susceptibility.[21–24]

Another hypothesis is that "mild mutations" of the APC or MMR genes might give rise to a milder form of cancer predisposition. As noted above, only mutations resulting in protein truncation are usually sufficient to cause FAP and probably HNPCC. Less severe mutations resulting in partially functional rather than truncated proteins might represent such a phenomenon. In fact, an attenuated form of FAP has now been observed, called attenuated APC (AAPC).[25] Persons with this condition have an average of 30 colonic adenomas, compared to the average of 1000 observed in fully developed typical FAP. The emergence of polyps and cancer is also delayed approximately 10 years in AAPC. Interestingly, the APC mutations in AAPC families found to date are truncating mutations, but are found at an extreme proximal or distal location of the gene. Another example of an APC mutation resulting in a milder phenotype was recently found in the Ashkenazi Jewish population.[26] The particular mutation was found in 6% of Ashkenazi Jewish people, but in almost 12% of Ashkenazi Jews with colon cancer and 28% of that population with colon cancer and a positive family history of large bowel malignancy. The investigators thus suggested an almost 2-fold risk attributable to this mutation, although the risk may be somewhat less than this. Elevated serum gastrin levels have been associated with an increased risk of colon cancer according to a recent study.[27] The study suggested that elevated gastrin could account for as many as 8% of cases of colon cancer. Chronically elevated serum gastrin levels are often found in the setting of atrophic gastritis, a condition that may arise from inherited predisposition and/or certain chronic environmental exposures.

SUMMARY

Familial risk of colon cancer is a commonly encountered issue, involving both rare inherited syndromes of colon cancer and common familial clustering of cases. The genes that give rise to the rare syndromes of FAP and HNPCC are now known and current research is addressing the cellular mechanisms of these genes and the proper application of genetic testing in families with one of the syndromes. Common familial clustering of cases appears to arise from an interaction of mildly to moderately severe inherited susceptibility factors with certain environmental factors to give rise to adenomatous polyps and then finally colorectal cancer. Research in this area involves identification of these purportedly more common susceptibility genes, and determination of how each interacts with environmental factors to give rise to polyps and cancer. Optimal application of varying degrees of familial risk to screening strategies is also being determined.

REFERENCES

1. Burt, R.W. Polyposis syndromes. *In*: Yamada, T. (ed): Textbook of Gastroenterology, 2nd edition. Philadelphia. 1995, pp. 1944–1966, J.B. Lippincott.
2. Morin, P.J., Sparks, A.B., Korinek, V., et al. Activation of β-catenin-Tcf signaling in colon cancer by mutations in β-catenin and *APC*. *Science* 275:1787–1790, 1997.
3. Nagase, H., Miyoshi, Y., Horii, A., et al. Correlation between the location of germ-line mutations in the *APC* gene and the number of colorectal polyps in familial adenomatous polyposis patients. *Cancer Res.* 52:4055, 1992.
4. Giardiello, F.M., Brensinger, J.D., Petersen, G.M., et al. The use and interpretation of commercial *APC* gene testing for familial adenomatous polyposis. *N. Engl. J. Med.* 336:823–827, 1997.
5. Marra, G. and Boland, C.R. Hereditary nonpolyposis colorectal cancer: the syndrome, the genes, and historical perspectives. *J. Natl. Cancer. Inst.* 87:1114–1125, 1995.
6. Lynch, H.T., Smyrk, T.C., Watson, P., et al. Genetics, natural history, tumor spectrum, and pathology of hereditary nonpolyposis colorectal cancer: an updated review. *Gastroenterol.* 104:1535–1549, 1993.
7. Vasen, H.F.A., Wijnen, J.T., Menko, F.H., et al. Cancer risk in families with hereditary nonpolyposis colorectal cancer diagnosed by mutation analysis. *Gastroenterol* 110:1020–1027, 1996.
8. Marra, G. and Boland, R. DNA repair and colorectal cancer. *Gastroenterol. Clin. North. Am.* 25:755–772, 1996.
9. Wijnen, J., Khan. P.M., Vasen, H., et al. Hereditary nonpolyposis colorectal cancer families not complying with the Amsterdam criteria show extremely low frequency of mismatch-repair-gene mutations. *Am. J. Hum. Genet.* 61:329–335, 1997.
10. Wijnen, J.T., Vasen, H.F.A., Khan, M., et al. Clinical findings with implications for genetic testing in families with clustering of colorectal cancer. *N. Engl. J. Med.* 339:511–518, 1998.
11. Burt, R.W. Familial risk and colorectal cancer. *Gastroenterol. Clin. North. Am.* 25:793–803, 1996.
12. Burt, R.W., Bishop, D.T., Cannon, L.A., et al. Dominant inheritance of adenomatous colonic polyps and colorectal cancer. *N. Engl. J. Med.* 312:1540–1544, 1985.
13. Cannon-Albright, L.A., Skolnick, M.K., Bishop, D.T., et al. Common inheritance of susceptibility to colonic adenomatous polyps and associated colorectal cancers. *N. Engl. J. Med.* 319:533–537, 1988.

14. St. John, J.B., McDermott, F.T., Hopper, J.L., et al. Cancer risk in relatives of patients with common colorectal cancer. *Ann. Intern. Med.* 118:785–790, 1993.
15. Fuchs, C.S., Giovannucci, E.L., Colditz, G.A., et al. A prospective study of family history and the risk of colorectal cancer. *N. Engl. J. Med.* 331:1669–1674, 1994.
16. Ahsan, H., Neugut, A.I., Gargowski, G.C., et al. Family history of colorectal adenomatous polyps and increased risk for colorectal cancer. *Ann. Intern. Med.* 128:900–905, 1998.
17. Kerber, R.A., Slattery, M.L., Potter, J.D., et al. Risk of colon cancer associated with a family history of cancer or colorectal polyps: the diet, activity, and reproduction in colon cancer study. *Int. J. Cancer* 78:157–160, 1998.
18. Winawer, S.J., Zauber, A.G., Gerdes, H., et al. Risk of colorectal cancer in the families of patients with adenomatous polyps. *N. Engl. J. Med.* 334:82–87, 1996.
19. Pariente, A., Milan, C., Lafon, J., et al. Colonoscopic screening in first-degree relatives of patients with sporadic colorectal cancer: a case-control study. *Gastroenterol.* 115:7–12, 1998.
20. Slattery, M.L. and Kerber, R.A. Family history of cancer and colon cancer risk: the Utah population database. *J. Natl. Cancer Inst.* 86:1618–1626, 1994.
21. Lang, N.P., Butler, M.A., Massengill, J., et al. Rapid metabolic phenotypes for acetyltransferase and cytochrome P4501A2 and putative exposure to food-borne heterocyclic amines increase the risk for colorectal cancer or polyps. *Cancer Epidemiol. Biomarkers Prev.* 3:675–682, 1994.
22. Zhong, S., Wyllie, A.H., Barnes, D., et al. Relationship between the GSTM1 genetic polymorphism and susceptibility to bladder, breast, and colon cancer. *Carcinogenesis* 14:1821–1824, 1993.
23. Chenevix-Trench, G., Young, J., Coggan, M., et al. Glutathione S-transferase M1 and T1 polymorphisms: susceptibility to colon cancer and age of onset. *Carcinogenesis* 16:1655–1657, 1995.
24. Strange, R.C., Matharoo, B., Faulser, G.C., et al. The human glutathione S-transferases: a case-control study of the incidence of the GST1 0 phenotype in patients with adenocarcinoma. *Carcinogenesis* 12:25–28, 1991.
25. Spirio, L., Olschwang, S., Groden, J., et al. Alleles of the *APC* gene: an attenuated form of familial polyposis. *Cell* 75:951–957, 1993.
26. Laken, S.J., Petersen, G.M., Gruber, S.B., et al. Familial colorectal cancer in Ashkenazim due to a hypermutable tract in *APC*. *Nature Genetics* 17:79–83, 1997.
27. Thorburn, C.M., Friedman, G.D., Dickinson, C.J., et al. Gastrin and colorectal cancer: a prospective study. *Gastroenterol* 115:275–280, 1998.

12

COLONIC CELL PROLIFERATION AND APOPTOSIS IN RODENT SPECIES

Modulation by Diet

Robert S. Chapkin and Joanne R. Lupton

Faculty of Nutrition
Molecular and Cell Biology Group
Texas A&M University
College Station, Texas 77843-2471

1. THE ROLE OF APOPTOSIS IN COLON CANCER

Apoptosis is a form of programmed cell death shown to play a central role in normal colonic development.[1,2] During apoptosis, the nucleus and cytoplasm condense and the dying cell fragments form membrane bound apoptotic bodies which are subsequently digested by phagocytic cells.[3] In contrast to necrosis, this process avoids the release of noxious cellular contents, preventing the induction of an inflammatory response.

The colonic epithelium is a self renewing tissue where the rate of new cell production is equaled by the rate of cell loss.[1] In the rat colon, distal colonic epithelial cells are derived from a stem cell population at the base of the crypt and migrate from a region of active cell proliferation in the bottom two-thirds of the crypt towards the top of the crypt, obtaining an increasingly differentiated or apoptotic phenotype.[2] In contrast, in the proximal colon, proliferating cells are located in midcrypt and migrate bidirectionally up toward the luminal surface and down toward the crypt base.[4] Interestingly, as a result of stem cell division, the dimensions of crypts increase with age,[5] consistent with an increase in cell proliferation.[5–7] Exfoliation of colonocytes from the luminal surface of the crypt may be initiated by a passive sloughing process, expulsion by mechanical forces from neighboring cells, or the direct result of apoptosis. The balance between proliferation and apoptosis is critical to the maintenance of a steady-state number for cell populations in the colon. In general, dysregulation of this mechanism can disrupt homeostasis, resulting in clonal expansion, with the resultant

Colon Cancer Prevention, edited under the auspices of AICR
Kluwer Academic / Plenum Publishers, New York, 1999.

over-production of affected cells.[8] The programmed induction of cell death also represents a mechanism by which colonocytes possessing DNA damage can be deleted.[9] This protective process can occur anywhere along the crypt axis.[9] Therefore, when rates of colonic cell proliferation are increased and apoptosis is defective, attenuated or inactivated, the risk of DNA damage perpetuation and hence carcinogenic risk is increased.[2,8–11]

Until recently, colon cancer development was thought to primarily occur through increased cell proliferation at several stages of the tumorigenic process, and little emphasis was placed on apoptosis. The emphasis has now shifted.[12] It has been clearly established that the transformation of colonic epithelium to carcinoma is in part associated with a progressive inhibition of apoptosis.[8,9,11,13,14]

Using a variety of chemical carcinogens in mice, it was recently proposed that the differential cancer incidence in the small versus large intestine, can be attributed to the differential removal of mutated cells.[2] Damaged cells in the small intestine are removed by apoptosis, which removes undesirable genetic mutations. In contrast, in the colon, apoptosis may not be initiated as strongly in the stem cell region. The lower level of selective or targeted deletion may result in the perpetuation of deleterious mutations in the colonic stem cell population, possibly contributing to the higher incidence of cancers in this region of the intestine.[2] However, this hypothesis is likely overly simplistic, since quantitation of apoptosis in the irradiated rat shows clearly that temporal alterations of apoptotic indices in small and large intestine are similar.[15]

2. EXPERIMENTAL COLON CANCER MODELS

The striking distribution of proliferating and apoptotic cells in colonic crypts suggests that colonocyte-colonocyte and/or colonocyte-stroma interactions may be contributing to cell cycle control.[16] Clearly, the topological organization of cytokinetic modulators within or contiguous to the crypt cannot readily be examined using cell culture systems. Therefore, investigators have used a multitude of rodent models to investigate the processes that regulate proliferative status and induction of apoptosis in the colon (Table 1).

We consider the azoxymethane-injected rat to be a highly relevant colon tumorigenesis model system. Although the specific chemical etiology of colon and other diet-associated cancers remains unclear, the chemical carcinogen-induced rat colon tumor system is a valuable model for studying the interaction between mediators of colonic cell proliferation, differentiation, and apoptosis. Evidence to support use of this model includes: (i) utilization of azoxymethane (AOM) provides a clear distinction between tumor initiation and promotion; (ii) carcinogen-induced *ras* and *Apc* mutations occur as early events in colon carcinogenesis, similar to humans; (iii) the development of tumors is responsive to the amount and type of dietary fat, and (iv) AOM-induced colon tumors closely parallel human colonic neoplasia in pathologic features.[17–21] In addition, the similarity between the expression of intracellular modulators of apoptosis (β-catenin, Cox-II, *bcl*-2, *bax*, and *bcl*-X_L) in the AOM injected rat and in humans suggests that this experimental system is a good model for studying the mechanisms which regulate apoptosis in humans.[22–24] We have utilized this model system in order to demonstrate that a decrease in apoptosis, rather than an increase in cell proliferation, is a better predictor of colon tumorigenesis.[13]

Table 1. *In vivo* rodent models for studying the relationship between colonic cell proliferation, differentiation, apoptosis, and tumorigenesis

1. Rat/mouse, azoxymethane or dimethylhydrazine-injected[13]
2. Rat, diet restriction/aging[7]
3. Mouse/rat ischemia-reperfusion injury[83]
4. Mouse/rat fast-refeed-circadian rhythm[84]
5. Gamma-irradiated mouse[16]
6. Germ-free mouse[85]
7. Dextran sulfate sodium treated mouse[86]
8. Guinea pig[40]
9. p53 null mouse[14]
10. min/Apc$^{\Delta 716}$ mouse[87]
11. ApcMin/+ p53+/– mouse[88]
12. Cox II null mouse[56]
13. Bcl-2 null mouse[38]
14. Apc-chimeric transgenic mouse[49]
15. Forced expression of SV40 TAg^{Wt} in mouse intestine[26]
16. Forced expression of K-ras^{Val12} × SV40 TAg^{Wt} in mouse intestine[89]
17. Mismatch repair (Msh2) null mouse[90,91]
18. Dpc4 (Smad 4)/TGFβ signaling null mouse[92]
19. MGMT (O^6-methylguanine-DNA methyltransferase) transgenic mouse[93]
20. DCC null mouse[94]
21. p21$^{CIP/WAF1}$ null mouse[95]
22. Forced expression of E-cadherin mouse[47]
23. Dominant negative N-cadherin mutant mouse[48]

3. INTRACELLULAR MEDIATORS OF COLONIC CYTOKINETICS

Three phases of apoptosis can be arbitrarily defined.[25] The induction phase corresponds to the initiation of the apoptotic signal. This is followed by the effector phase, where the numerous apoptotic-promoting signals converge into a few pathways involving protease signaling of cell death. Finally, a degradation phase occurs where cellular structures and functions are destroyed. Recently, investigators have focused on the mechanisms responsible for the reduction of apoptosis in colon cancer by using a variety of *in vivo* rodent model systems.[10,26–28] Some of the molecular events which regulate colonic cytokinetics are described below. The distribution of pro and anti-apoptotic mediators in the colonic crypt is shown in Fig. 1.

3.1. Intestinal p53 and WAF1/Cip1

Analysis of the progressive accumulation of the genetic alterations that occur in colorectal tumorigenesis indicates that the pathways of colonic cell proliferation and cell death may be tightly coupled.[29] Evidence of this association is supported by the fact that a set of genes involved in cell cycle regulation also appears to act as tumor suppressor genes in cell death.[29,30] Among them, wild-type p53 can induce apoptosis and inhibit cell proliferation by implementing a G_1 arrest, while mutant p53 can inhibit apoptosis.[31] It is generally thought that DNA damage results in a post-translational increase in p53, which subsequently acts as a sequence-specific DNA-binding transcription factor.[31] WAF1/Cip1 is an example of a gene transcribed in a p53-dependent

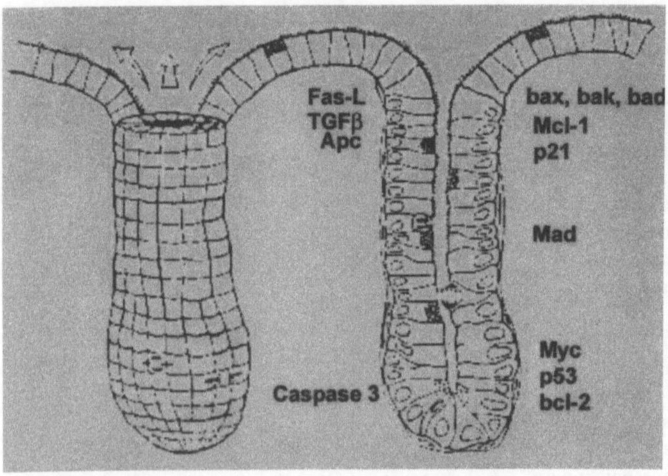

Figure 1. Localization of apoptotic mediators in the colonic crypt.

manner. WAF1/Cip1 inhibits G_1 cyclin-dependent kinases leading to transient cell cycle arrest, and is subject to precise topological control in the colon.[32,33] p53 is therefore necessary for a G_1 checkpoint that monitors damaged DNA. Presumably this allows time for the cell to effect DNA repair.[33] This is noteworthy because the loss or inactivation of p53 (mutant p53) is the most common single lesion in human colorectal cancer.[29,34] Mutant p53 is much more stable than wild type p53, accumulates in neoplastic cells and, thus, becomes immunologically detectable.[31] Therefore, positive immunostaining is indicative of abnormalities of the p53 gene and neoplasia.[31,34,35]

3.2. Crypt Levels of *bcl*-2 and *bax*

Malignant transformed cells can result from the progressive loss of apoptotic signals.[29,31] This is supported by findings that expression of oncogenes such as *bcl*-2 or mutant tumor suppressor genes such as mutant p53 (described above) may protect a cell from apoptosis.[10] Cells expressing either *bcl*-2 or mutant p53 are resistant to inducers of apoptosis.[8] *bcl*-2 acts specifically by blocking apoptotic cell death, and is inversely correlated with wild-type p53 expression in adenomas.[8] In the normal colon, *bcl*-2 expression is restricted to the lower half of crypts (refer to Fig. 1).[10,36–38] However, in colorectal carcinomas, *bcl*-2 is expressed along the entire crypt axis.[8,10] Recent data indicate that the abnormal activation of the *bcl*-2 gene is an early event in the malignant transformation of colonic epithelium and may facilitate tumor progression.[8,9] In contrast, *bax*, which heterodimerizes with *bcl*-2 related proteins and promotes cell death, is localized in the apical colonic epithelium (refer to Fig. 1).[39] Additional evidence indicating that the induction of apoptosis is accompanied by the enhanced expression of *bax* has been recently demonstrated.[40] Using tissue sheets from guinea pig proximal colon, it was demonstrated that the removal of butyrate from Ussing chambers was associated with the induction of *bax* which paralleled a rapid process of apoptosis ex vivo.[40] This is consistant with the concept that the ratio of *bcl*-2/*bax* can determine whether a given cell will execute or ignore an apoptotic stimulus.[41] To date, the expression and localization of *bcl*-2 and its 11 related genes (antagonists of apoptosis, *bcl*-2, *bcl*-x_L, *bcl*-w, and *bfl*-1, BRAG-1, and *mcl*-1; agonists of apoptosis, *bax*, *bik*, *bad*, *bak*, and *bcl*-x_s) in the colon following carcinogen adminstration and diet manipulation have not been determined.

4. SPONTANEOUS AND DAMAGE-INDUCED APOPTOSIS

Programmed cell death generally (although not exclusively) occurs after terminal differentiation.[42] However, unlike the small intestine, spontaneous apoptosis in the proliferative zone of the colon is a rare event.[16] Therefore, spontaneous apoptosis in the colon is unlikely to effectively regulate stem cell numbers. Although, similar to the small intestine, it may be an effective means of removing senescent cells from the luminal surface of the crypt.[42] An important exception to this sequence of events exists during the initiation of tumorigenesis when there is an immediate apoptotic response to DNA damage induced by either carcinogen treatment[43] or γ-radiation.[38] This form of apoptosis is targeted primarily to actively proliferating cells that have undergone DNA damage. Despite this important exception, under normal homeostatic conditions, the typical pattern of colonic epithelial cytokinetics is for stem cells to divide and produce daughter cells, which migrate up the crypt column, undergo several *in transit* divisions, differentiate in the upper one-third of the crypt, undergo apoptosis, and are exfoliated and removed in the fecal stream.[16] The sequence of events from terminal differentiation to apoptosis is unknown. Whether the differentiation of colonocytes triggers the initiation of apoptosis or is merely a necessary prerequisite for the apoptotic event, which is initiated by other signals, remains to be determined. Cancer cells usually have a less differentiated phenotype than their normal counterparts,[44] and in most tumor cell lines, a more differentiated phenotype is considered to be less metastatic. Therefore, differentiating agents are thought to be antineoplastic in part because they predispose the cell to undergo apoptosis.[44]

A number of investigators using the p53, null mouse have demonstrated that spontaneous apoptosis in the small intestine and colon is largely a p53-independent process.[14,15,45] In contrast, targeted apoptosis (radiation damage-induced) is p53 dependent. In *bcl*-2 null mice, there is an increase in the levels of spontaneous apoptosis in cells at the base of the colonic crypt.[38] No such increase in spontaneous apoptosis is observed in small intestinal crypts. The pattern was similar following exposure to damaging radiation. These data indicate that *bcl*-2 plays an important role in attenuating both spontaneous and damage-induced (targeted) apoptosis in the colon. Future research examining the interaction of *bcl*-2, p53, and *bax* in a series of doubly mutant mice should help elucidate the homeostatic controls on colonic crypt cytokinetics.

5. CELL ANCHORAGE REGULATES APOPTOSIS

Epithelial cells in the crypt are dependent upon adhesion to extracellular matrix for survival. Loss of contact results in apoptosis, also termed anoikis.[46] The role of specific cell surface adhesion molecules, e.g., Apc and cadherins, has recently been examined in a series of experiments using chimeric transgenic mice exhibiting forced intestinal expression of E-cadherin,[47] dominant negative N-cadherin mutation,[48] or Apc^{WT}.[49] These studies demonstrate that the balance between cell proliferation and apoptosis is in part coordinated by cell-cell interactions. In addition, β-catenin, a key component of the cadherin-mediated cell-cell adhesion system, is frequently mutated in azoxymethane-induced rat colon tumors.[24] This is relevant because β-catenin is an oncogene, and Apc acts as a negative regulator of β-catenin signaling.[49] Indeed, the regulation of β-catenin is critical to Apc's tumor suppressive effect. Disruption by mutations in either Apc or β-catenin may reduce cell migration, stimulate proliferation and antagonize apoptosis in a dominant negative fashion.[51,52] Collectively, these events result in anoikis resistance. It is believed that the breakdown of normal cell-cell communication is associated with colon carcinogenesis.

6. EXPRESSION OF CYCLOOXYGENASES (CONSTITUTIVE COX-I AND INDUCIBLE COX-II) IN COLON CANCER

Nonsteroidal antiinflammatory drugs (NSAIDs) are strong inhibitors of tumor formation in rodent models of chemically-induced colon cancer.[53] NSAIDs cause a downward shift in the cell proliferation compartment of colonic crypts,[54] and appear to increase apoptosis.[55] There are compelling data that the protective effects of NSAIDs are in part related to the inhibition of intestinal Cox-II, since the induction of Cox-II is an early, rate-limiting step for adenoma formation.[56] Elevated Cox-II expression in the colon has also been associated with increased levels of *bcl*-2 and E-cadherin, resulting in increased adhesion to extracellular matrix.[57] This could account for the anti-apoptotic effect observed. However, the NSAID-dependent inhibition of colonic Cox-I as a contributing factor cannot be discounted, since Cox-I promotes crypt stem cell survival and proliferation.[58] Recently, Chan et al.[59] have shown that the tumor suppressive effects of NSAIDs are not likely related to a reduction in Cox I/II-derived prostaglandins but to an elevation of the prostaglandin precursor, arachidonic acid (20:4n–6). Using colonic cell lines, they demonstrated that the NSAID-induced increase in arachidonic acid stimulates the conversion of sphingomyelin to ceramide, a known mediator of apoptosis. We have been unable to confirm the presence of a colonic arachidonate-ceramide-apoptosis signaling cascade *in vivo*.[60]

7. DIETARY MODULATION OF COLONIC CYTOKINETICS

The two dietary components thought to have the most significant effect on colon tumor development are dietary fat and fiber. There are considerable data to support the concept that the type of fat or fiber is actually more important to tumor development than is the amount of either of these components in the diet.[61–64] Specifically, n-3 polyunsaturated fatty acids (n-3 PUFA) are protective against colon cancer in experimental studies.[61,65,66] These fatty acids are found in high concentration in fish and fish oils. The type of dietary fiber that is the most protective against colon tumor development remains the subject of debate. Butyrate, a short chain fatty acid derived in the colon from microbial fermentation, can act as an inducer of apoptosis in colonic epithelial cell lines.[67,68] These *in vitro* findings have resulted in the hypothesis that the more fermentable fibers should be the most protective against colon cancer since they produce the highest amounts of butyrate *in vivo*.[69] However, there is some discrepancy between *in vitro* and *in vivo* studies regarding the efficacy of butyrate[64,69,70] and this issue is in need of resolution.

Interestingly, caloric restriction in aging rats is capable of increasing colonic apoptosis.[7] Therefore, caloric restriction may not only reduce colon cancer risk by lowering the rate of epithelial cell proliferation,[71] but may also protect the colon from accumulation of DNA-altered cells during the aging process by enhancing the deletion of damaged cells.

8. INTERACTIVE EFFECT OF FAT AND FIBER ON APOPTOSIS DURING COLON TUMOR DEVELOPMENT

In the well documented rat model of colon carcinogenesis, a plethora of studies find that fish oil is protective against colon tumor development compared to corn oil based diets (containing primarily n-6 PUFA).[61,65,72] In addition, this protective effect of fish oil is enhanced when highly-fermentable pectin, rather than poorly fermentable cellulose, is in the diet (refer to Fig. 2).[61,72] We also find that fish oil results in a greater degree of colonocyte differentiation, and apoptosis compared to corn oil feeding.[61] These effects are enhanced in the fish oil/pectin diet treatment group (Fig. 3). Fish oil/pectin, compared to corn oil/pectin, is also protective at the very earliest stages of colon carcinogenesis (within 12 h of carcinogen injection).[74] During this initiation phase of tumor induction, fish oil/pectin stimulates apoptosis in the upper portion of the crypt and down-regulates DNA damage. It is important to note that tumorigenesis begins in this upper portion of the crypt, initially expanding into the colonic lumen (as a polyp) and eventually extending downwards into the muscularis mucosa, finally becoming a carcinoma. It is clear from these studies that fish oil is protective against experimentally-induced colon cancer by upregulating apoptosis and that this protective effect is enhanced by co-administration of dietary pectin. Although the fish oil versus corn oil differences are significant, they are numerically small and one might

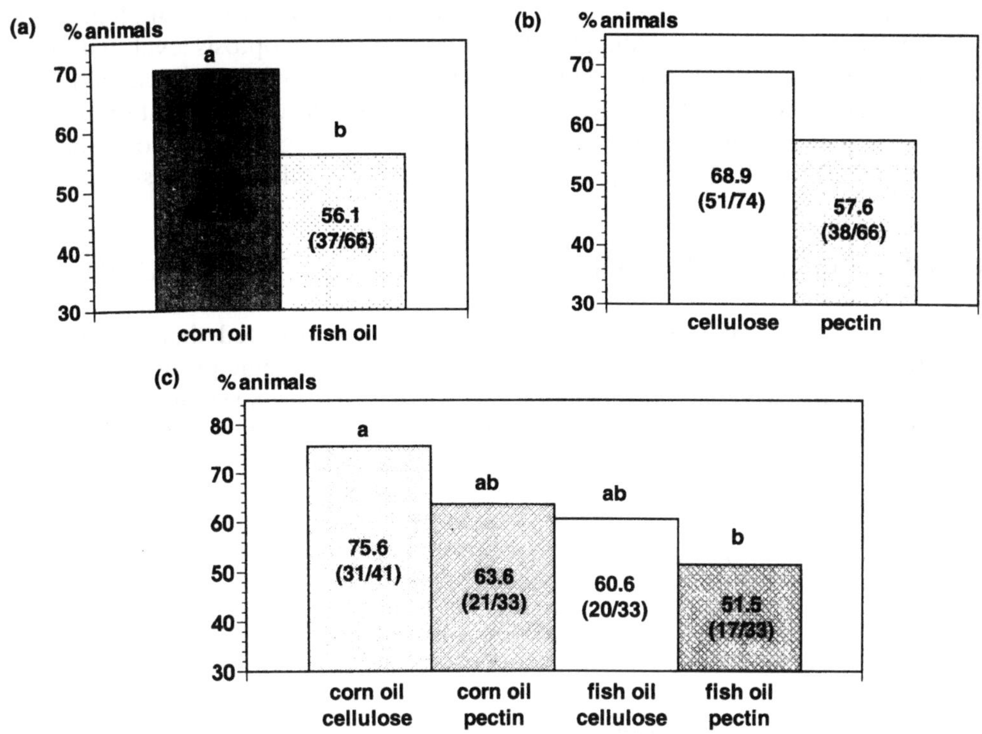

Figure 2. Colon adenocarcinomas incidence. Dietary fish oil and pectin synergize to protect against colon tumor development. Rats were fed semi-purified diets containing 15% by weight fish oil or corn oil and terminated 36 weeks following the second azoxymethane injection. No tumors were observed in the saline-injected (control) group. Adapted from Chang.[13]

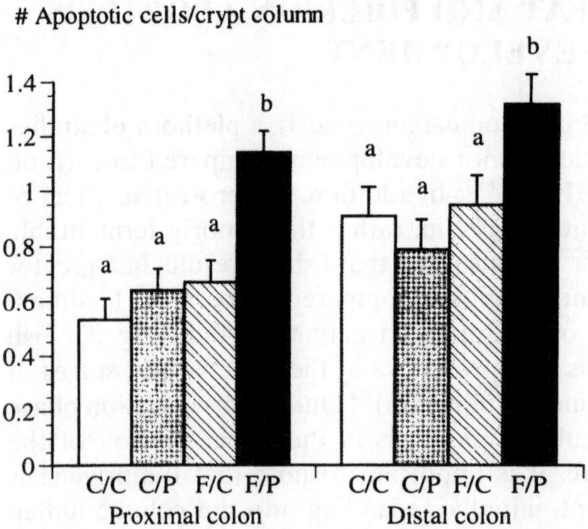

Figure 3. The interactive effect of dietary fat and fiber on the number of apoptotic cells per crypt column in rat proximal and distal colon. Rats fed fish oil/pectin, F/P, had more apoptotic cells per crypt column compared to rats fed corn oil/cellulose, C/C; corn oil/pectin, C/P; or fish oil/cellulose, F/C. Adapted from Chang.[61]

question their biological relevance. However, a single extra stem cell, not deleted by apoptosis, will produce 60–120 extra cells per crypt.[16] Small changes in apoptosis in this area, therefore, as a function of diet, may have important biological consequences, particularly if this enhanced apoptosis seen with fish oil is targeted to DNA damaged cells. Importantly, we found a uniquely different response of fish oil supplemented animals to DNA damage in this upper one-third of the crypt compared to corn oil supplemented animals.[74] Specifically, as DNA adduct levels increased, so did apoptosis levels with fish oil feeding. In contrast, as DNA adduct levels increased with corn oil supplementation, apoptosis actually decreased (Fig. 4). This suggests that fish oil supplementation results in the targeting of apoptosis to DNA damaged cells. This targeted apoptosis should decrease DNA adducted cells and provide less possibility for clonal expansion.

9. REGULATION OF APOPTOSIS BY PROTEIN KINASE C (PKC)

Using the AOM-rat tumor model, we recently demonstrated that n-3 PUFA prevent the carcinogen-induced chronic down-regulation of colonic protein kinase C (PKC) δ, ζ, and the selective up-regulation of PKC β_{II}.[75,76] This is significant because the maintenance of crypt PKC isozyme levels may sustain the homeostatic balance

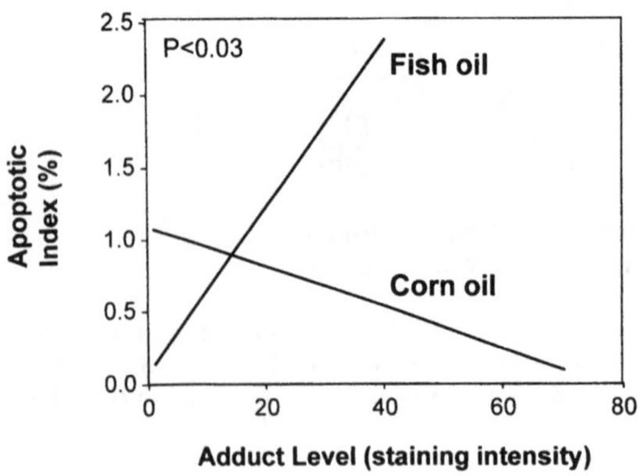

Figure 4. Fish oil fed animals target DNA damaged cells for deletion by apoptosis. Adapted from Hong.[74] A positive slope (fish oil fed rats) indicates that an increase in apoptosis is associated with an increase in the level of DNA adducts.

between cell proliferation and apoptosis.[77–80] Indeed, the cytoskeletal localization of select PKC isoforms suggests that colonic PKC may be an important regulator of crypt cell-cell communication.[81] The ability of n-3 PUFA to block the carcinogen-induced increase in PKC β_{II} expression is consistent with previous reports citing a role for PKC β_{II} in colon tumorigenesis.[76,78,82] Therefore, it is possible that dietary n-3 PUFA reduce colon cancer incidence in part by blocking the effects of carcinogen on colonic PKC isozyme-related signal transduction.

10. CONCLUSION

The colonic epithelium is a self-renewing tissue in which the rate of new cell production is equaled by the rate of cell loss. The balance between proliferation and apoptosis is critical to the maintenance of steady-state number for cell populations in the colon. Recently, the universal applicability of changes in cell proliferation as a diagnostic marker for colonic tumor development has been questioned. Experimental studies in rodent species have shown that changes in colonic cell proliferation do not always predict tumor development. This likely indicates that the relative contribution of an up-regulation of cell division to the tumorigenic process is episodic in nature. It has now been clearly established that the transformation of colonic epithelium to carcinomas is associated with a progressive inhibition of apoptosis. Interestingly, in the rat azoxymethane colon tumorigenic model, we have demonstrated that apoptosis has greater prognostic value compared to cell proliferation in terms of predicting which animals will develop colonic tumors. These data suggest that assessment of apoptosis deserves a place in the armamentarium of intermediate biomarkers for colon tumorigenesis.

It is now well accepted that colon cancer evolves from a multi-step process and is a disease strongly influenced by environmental factors, with diet being one of the most important modifying agents. Among dietary factors, there are strong experimental data using a variety of rat and mouse colon carcinogenesis models indicating a protective effect of select fats and fibers. We have recently demonstrated that the balance between colonic epithelial cell proliferation and apoptosis in the rat colon can be favorably modulated by feeding a combination of n-3 PUFA (found in fish oil) and pectin (a highly fermentable fiber), thereby conferring resistance to toxic carcinogenic agents. These data offer insight into the mechanisms by which select dietary factors modify the risk of colon cancer development. Elucidation of the mechanisms by which select dietary fats and fibers reduce colon cancer incidence will lead to the establishment of dietary guidelines designed to reduce colon cancer morbidity and mortality.

ACKNOWLEDGMENT

This research was supported by the National Institutes of Health (CA59034 and CA61750) and the American Institute for Cancer Research (97B-020).

REFERENCES

1. Podolsky, D.K. Regulation of intestinal epithelial proliferation: a few answers, many questions. *Am. J. Physiol.* 264:G179–G186, 1993.

2. Potten, C.S., Li, Y.Q., O'Connor, P.J., and Winton, D.J. A possible explanation for the differential cancer incidence in the intestine, based on distribution of the cytotoxic effects of carcinogens in the murine large bowel. *Carcinogenesis* 13:2305–2312, 1992.
3. Thompson, C.B. Apoptosis in the pathogenesis and treatment of disease. *Science* 267:1456–1462, 1995.
4. Sato, M. and Ahnen, D.J. Regional variability of colonocyte growth and differentiation in the rat. *Anatom. Record* 233:409–414, 1992.
5. Martin, K., Kirkwood, T.B.L., and Potten, C.S. Age changes in stem cells of murine small intestinal crypts. *Exp. Cell Res.* 241:316–323, 1998.
6. Holt, P.R. and Yeh, K.Y. Colonic proliferation is increased in senescent rats. *Gastroenterol* 95: 1556–1563, 1988.
7. Holt, P.R., Moss, S.F., Heydari, A.R., and Richardson, A. Diet restriction increases apoptosis in the gut of aging rats. *J. Gerontol.* 53A:B168–B172, 1998.
8. Siniscope, F.A., Ruan, S.B., Cleary, K.R., Stephens, L.C., Lee, J.J., and Levin, B. *bcl*-2 and p53 oncoprotein expression during colorectal tumorigenesis. *Cancer Res.* 55:237–241, 1995.
9. Bedi, A., Pasricha, P.J., Alchtar, A.J., Barber, J.P., Bedi, G.C., Giairdiello, F.M., Zehnbauer, B.A., Hamilton, S.R., and Jones, R.J. Inhibition of apoptosis during development of colorectal cancer. *Cancer Res.* 55:1811–1816, 1995.
10. Hayashi, R., Luk., H., Horio, D., and Dashwood, R. Inhibition of apoptosis in colon tumors induced by 2-amino-3-methylimidazo[4,5-f]quinoline. *Cancer Res.* 56:4307–4310, 1996.
11. Preston-Martin, S., Pike, M.C., Ross, R.K., Jones, P.A., and Henderson, B.E. Increased cell division as a cause of human cancer. *Cancer Res.* 50:7415–7412, 1990.
12. Farber, E. Cell proliferation as a major risk factor for cancer: a concept of doubtful validity. *Cancer Res.* 55:3759–3762, 1995.
13. Chang, W.C.L., Chapkin, R.S., and Lupton, J.R. Predictive value of proliferation, differentiation and apoptosis as intermediate markers for colon tumorigenesis. *Carcinogenesis* 18:721–730, 1997.
14. Merritt, A.J., Potten, C.S., Kemp, C.J., Hickman, J.A., Balmain, A., Lane, D.P., and Hall, P.A. The role of p53 in spontaneous and radiation-induced apoptosis in the gastrointestinal tract of normal and p53-deficient mice. *Cancer Res.* 54:614–617, 1994.
15. Arai, T., Kida, Y., Harmon, B.V., and Gobe, G.C. Comparative alterations in p53 expression and apoptosis in the irradiated small and large intestine. *Br. J. Cancer* 74:406–412, 1996.
16. Potten, C.S., Wilson, J.W., and Booth, W.C. Regulation and significance of apoptosis in the stem cells of the gastrointestinal tract. *Stem Cells* 15:82–93, 1997.
17. Ahnen, D.J. Are animal models of colon cancer relevant to human diseases? *Dig. Dis. Sci.* 30:103S–106S, 1985.
18. Chapkin, R.S., Jiang, Y.H., Davidson, L.A., and Lupton, J.R. Modulation of intracellular second messengers by dietary fat during colonic tumor development. *Adv. Exp. Med. Biol.* 422:85–96, 1997.
19. Kakiuchi, H., Watanabe, M., Ushijima, T., Toyota, M., Imai, K., Weisburger, J.H., Sugimura, T., and Nagao, M. Specific 5′-GGGA-3′ → 5′-GGA-3′ mutation of the Apc gene in rat colon tumors induced by 2-amino-1-methyl-6-phenylimidazo[4,5-b]pyridine. *Proc. Natl. Acad. Sci. USA* 92:910–914, 1995.
20. Maltzman, T., Whittington, J., Driggers, L., Stephens, J., and Ahnen, D. AOM-induced mouse colon tumors do not express full-length APC protein. *Carcinogenesis* 18:2435–2439, 1997.
21. Reddy, B.S., Burill, C., and Rigotty, J. Effect of diets high in ϖ-3 and ϖ-6 fatty acids on initiation and postinitiation stages of colon carcinogenesis. *Cancer Res.* 51:487–491, 1991.
22. Dubois, R.N., Radhika, A., Reddy, B.S., and Entingh, A.J. Increased cyclooxygenase-2 levels in carcinogen-induced rat colonic tumors. *Gastroenterol.* 110:1259–1262, 1996.
23. Hirose, Y., Yoshimi, N., Suzui, M., Kawabata, K., Tanaka, T., and Mori, H. Expression of bcl-2, bax, and bcl-X_L proteins in axoxymethane-induced rat colonic adenocarcinomas. *Molec. Carcinogenesis* 19:25–30, 1997.
24. Takahashi, M., Fukuda, K., Sugimura, T., and Wakabayashi, K. β-Catenin is frequently mutated and demonstrates altered cellular location in azoxymethane-induced rat colon tumors. *Cancer Res.* 58:42–46, 1998.
25. Cory, S. and Adams, J.M. Matters of life and death: programmed cell death at Cold Spring Harbor. *Biochim. Biophys. Acta* 1377:R25–R44, 1998.
26. Chandrasekaran, C., Coopersmith, C.M., and Gordon, J.I. Use of normal and transgenic mice to examine the relationship between terminal differentiation of intestinal epithelial cells and accumulation of their cell cycle regulators. *J. Biol. Chem.* 271:28414–28421, 1996.

27. Rosenberg, M.P. Gene knockout and transgenic technologies in risk assessment: The next generation. *Molec. Carcinogenesis* 20:262–274, 1997.
28. Zhang, T., Nanney, L.B., Luongo, C., Lamps, L., Heppner, K.J., DuBois, R.N., and Beauchamp, R.D. Concurrent overexpression of cyclin D1 and cyclin-dependent kinase 4 (Cdk4) in intestinal adenomas from multiple intestinal neoplasia (Min) mice and human familial adenomatous polyposis patients. *Cancer Res.* 57:169–175, 1997.
29. Harrington, E.A., Fanidi, A., and Evan, G.I. Oncogenes and cell death. *Curr. Opin. Genet. Devel.* 4:120–129, 1994.
30. Gervais, J.L., Seth, P., and Zhang, H. Cleavage of CDK inhibitor $p21^{Cip/Waf1}$ by caspases is an early event during DNA damage-induced apoptosis. *J. Biol. Chem.* 273:19207–19212, 1998.
31. Picksley, S.M. and Lane, D.P. p53 and Rb: their cellular roles. *Curr. Opin. Cell Biol.* 6:853–858, 1994.
32. El-Deiry, W.S., Tokino, T., Waldman, T., Oliner, J.D., Velculescu, V.E., Burrell, M., Hill, D.E., Healy, E., Rees, J.L., Hamilton, S.R., Kinzler, K.W., and Vogelstein, B. Topological control of $p21^{WAF/CIP1}$ expression in normal and neoplastic tissues. *Cancer Res.* 55:2910–2919, 1995.
33. Peter, M. and Herskowitz, I. Joining the complex: cyclin-dependent kinase inhibitory proteins and the cell cycle. *Cell* 79:181–184, 1994.
34. Rodrigeues, N.R., Rowan, A., Smith, M.E.F., Kerr, I.B., Bodmer, W.F., Gannon, J.V., and Lane, D.P. p53 mutations in colorectal cancer. *Proc. Natl. Acad. Sci. USA* 87:7555–7559, 1990.
35. Starzynska, T., Bromley, M., Ghosh, A., and Stern, P.L. Prognostic significance of p53 overexpression in gastric and colorectal carcinoma. *Br. J. Cancer* 66:558–562, 1992.
36. Hague, A., Moorghen, M., Hicks, D., Chapman, M., and Paraskeva, C. BCL-2 expression in human colorectal adenomas and carcinomas. *Oncogene* 9:3367–3370, 1994.
37. Hockenberry, D.M., Zutter, M., Hickey, W., Nahm, M., and Korsmeyer, S.J. BCL-2 protein is topographically restricted in tissues characterized by apoptotic cell death. *Proc. Natl. Acad. Sci. USA* 88:6961–6965, 1991.
38. Merritt, A.J., Potten, C.S., Watson, A.J.M., Loh, D.Y., Nakayama, K., Nakayama, K., and Hickman, J.A. Differential expression of bcl-2 in intestinal epithelia. Correlation with attenuation of apoptosis in colonic crypts and the incidence of colonic neoplasia. *J. Cell Sci.* 108:2261–2271, 1995.
39. Krajewski, S., Krajewska, M., Shabaik, A., Miyashita, T., Wang, H.G., and Reed, J.C. Immunohistochemical determination of *in vivo* distribution of bax, a dominant inhibitor of bcl-2. *Am. J. Pathol.* 145:1323–1336, 1994.
40. Hass, R., Busche, R., Luciano, L., Reale, E., and Englehardt, W.V. Lack of butyrate is associated with induction of bax and subsequent apoptosis in the proximal colon of guinea pig. *Gastroenterol.* 112: 875–881, 1997.
41. Lee, J. M. and Bernstein, A. Apoptosis, cancer, and the p53 tumor suppressor gene. *Cancer Metast. Rev.* 14:149–161, 1995.
42. Hall, P.A., Coates, P.J., Ansari, B., and Hopwood, D. Regulation of cell number in the mammalian gastrointestinal tract: the importance of apoptosis. *J. Cell Sci.* 107:3569–3577, 1994.
43. Hirose, Y., Yoshimi, N., Makita, H., Hara, A., Tanaka, T., and Mori, H. Early alterations of apoptosis and cell proliferation in azoxymethane-initiated rat colonic epithelium. *Jpn. J. Cancer Res.* 87:575–582, 1996.
44. Beere, H.M. and Hickman, J.A. Differentiation: a suitable strategy for cancer chemotherapy? *Anti-Cancer Drug Design* 8:299–322, 1993.
45. Clarke, A.R., Gledhill, S., Hooper, M.L., Bird, C.C., and Wyllie, A.H. p53 dependence of early apoptotic and proliferative responses within the mouse intestinal epithelium following γ-irradiation. *Oncogene* 9:1767–1773, 1994.
46. Cardone, M.H., Salvesen, G.S., Widmann, C., Johnson, G., and Frisch, S.M. The regulation of anoikis: MEKK-1 activation requires cleavage by caspases. *Cell* 90:315–323, 1997.
47. Hermiston, M.L., Wong, M.H., and Gordon, J.I. Forced expression of E-cadherin in the mouse intestinal epithelium slows cell migration and provides evidence for nonautonomous regulation of cell fate in a self-renewing system. *Genes Development* 10:985–996, 1996.
48. Hermiston, M.L. and Gordon, J.I. Inflammatory bowel disease and adenomas in mice expressing a dominant negative N-cadherin. *Science* 270:1203–1207, 1995.
49. Wong, M.H., Hermiston, M.L., Syder, A.J., and Gordon, J.I. Forced expression of the tumor suppressor adenomatosis polyposis coli protein induces disordered cell migration in the intestinal epithelium. *Proc. Natl. Acad. Sci. USA* 93:9588–9593, 1996.
50. Morin, P.J., Sparks, A.B., Korinek, V., Barker, N., Clevers, H., Vogelstein, B., and Kinzler, K.W.

Activation of β-catenin-Tcf signaling in colon caner by mutations in β-catenin or Apc. *Science* 275:1787–1790, 1997.

51. Mahmoud, N.N., Boolbol, S.K., Bilinski, R.T., Martucci, C., Chadburn, A., and Bertagnolli, M.M. Apc gene mutation is associated with a dominant-negative effect upon intestinal cell migration. *Cancer Res.* 57:5045–5050, 1997.
52. Peifer, M. β-catenin as oncogene: the smoking gun. *Science* 275:1752–1753, 1997.
53. Reddy, B.S., Rao, C.V., Rivenson, A., and Kelloff, G. Inhibitory effect of aspirin on azoxymethane-induced colon carcinogenesis in F334 rats. *Carcinogenesis* 14:1493–1497, 1993.
54. Barnes, C.J., Lee, M., Hardmann, W.E., and Cameron, I.L. Aspirin, age, and proximity to lymphoid nodules influence cell proliferation parameters in rat colonic crypts. *Cell Prolif.* 28:59–71, 1995.
55. Samaha, H.S., Kelloff, G.J., Steele, V., Rao, C.V., and Reddy, B.S. Modulation of apoptosis by sulinda, curcumin, phenylethyl-3-methylcaffeate, and 6-phenylhexyl isothiocyanate: apoptotic index as a biomarker in colon cancer chemoprevention and promotion. *Cancer Res.* 57:1301–1305, 1997.
56. Oshima, M., Dinchuk, J.E., Kargman, S.L., Ohsima, H., Hancock, B., Kwong, E., Trzaskos, J.M., Evans, J.F., and Taketo, M.M. Suppression of intestinal polyposis in ApcΔ716 knockout mice by inhibition of cyclooxygenase 2 (COX-2). *Cell* 87:803–809, 1996.
57. Tsujii, M. and DuBois, R.N. Alterations in cellular adhesion and apoptosis in epithelial cells overexpresing prostaglandin endoperoxide synthase 2. *Cell* 83:493–501, 1995.
58. Cohn, S.M., Schloemann, S., Tessner, T., Seibert, K., and Stenson, W.F. Crypt stem cell survival in the mouse intestinal epithelium is regulated by prostaglandins synthesized through cyclooxygenase-1. *J. Clin. Invest.* 99:1367–1379, 1997.
59. Chan, T.A., Morin, P.J., Vogelstein, B., and Kinzler, K.W. Mechanisms underlying nonsteroidal anti-inflammatory drug-mediated apoptosis. *Proc. Natl. Acad. Sci. USA* 95:681–686, 1998.
60. Jiang, Y.H., Lupton, J.R., Jolly, C.A., Davidson, L.A., Aukema, H.M., and Chapkin, R.S. Dietary fat and fiber differentially regulate intracellular second messengers during tumor development in rat colon. *Carcinogenesis* 17:1227–1233, 1996.
61. Chang, W.C.L., Chapkin, R.S., and Lupton, J.R. Fish oil blocks azoxymethane-induced tumorigenesis by increased cell differentiation, and apoptosis rather than decreased cell proliferation. *J. Nutr.* 128:491–497, 1998.
62. Kim, D.Y., Chung, K.H., and Lee, J.H. Stimulatory effects of high-fat diets on colon cell proliferation depend on the type of dietary fat and site of the colon. *Nutr. Cancer* 30:118–123, 1998.
63. Hong, M.Y., Chang, W.C., Chapkin, R.S., and Lupton, J.R. Relationship among colonocyte proliferation, differentiation, and apoptosis as a function of diet and carcinogen. *Nutr. Cancer* 28:20–29, 1997.
64. Zoran, D.L., Turner, N.D., Taddeo, S.S., Chapkin, R.S., and Lupton, J.R. Wheat bran reduces tumor incidence in a rat model of colon cancer independent of effects on distal luminal butyrate concentrations. *J. Nutr.* 127:2217–2225, 1997.
65. Reddy, B.S. Chemoprevention of colon cancer by dietary fatty acids. *Cancer Metastasis Rev.* 13:285–302, 1994.
66. Takahashi, M., Fukutake, M., Isoi, T., Fukada, K., Sato, H., Yazawa, K., Sugimura, T., and Wakabayashi, K. Suppression of azoxymethane-induced rat colon carcinoma development by a fish oil component, docosahexaenoic acid (DHA). *Carcinogenesis* 18:1337–1342, 1997.
67. Singh, B., Halestrap, P., and Paraskeva, C. Butyrate can act as a stimulator of growth or inducer of apoptosis in human colonic cell lines depending on the presence of alternative energy sources. *Carcinogenesis* 18:1265–1270, 1997.
68. Williams, A.C., Hague, A., Elder, D.J.E., and Paraskeva, C. In vitro models for studying colorectal carcinogenesis cellular and molecular events including APC and Rb cleavage in the control of proliferation, differentiation, and apoptosis. *Biochim. Biophys. Acta* 1288:F9–F19, 1996.
69. Lupton, J.R. Butyrate and colonic cytokinetics: differences between *in vitro* and *in vivo* studies. *Eur. J. Cancer Prev.* 4:373–378, 1995.
70. Caderni, G., Luceri, C., Lancioni, L., Tessitore, L., and Dolara, P. Slow-release pellets of sodium butyrate increase apoptosis in the colon of rats treated with azoxymethane, without affecting aberrant crypt foci and colonic proliferation. *Nutr. Cancer* 30:175–181, 1998.
71. Steinbach, G., Kumar, S.P., Reddy, B.S., Lipkin, M., and Holt, P.R. Effects of caloric restriction and dietary fat on epithelial cell proliferation in rat colon. *Cancer Res.* 53:2745–2749, 1993.
72. Deschner, E.E., Lytle, J., Wong, G., Ruperto, J., and Newmark, H.L. The effect of dietary omega-3 fatty acids (fish oil) on azoxymethanol-induced focal areas of dysplasia and colon tumor incidence. *Cancer* 66:2350–2356, 1990.

73. Lee, D.Y.K., Chapkin, R.S., and Lupton, J.R. Dietary fat and fiber modulate colonic cell proliferation in an interactive site specific manner. *Nutr. Cancer* 20:107–118, 1993.
74. Hong, M.Y., Chapkin, R.S., Turner, N.D., Galindo, C.D., Carroll, R.J., and Lupton, J.R. Fish oil enhances targeted apoptosis of colonocytes within the first 12 hours of carcinogen exposure and results in lower levels of DNA damage compared to corn oil. *FASEB J.* 12:A656, 1998.
75. Davidson, L.A., Aymond, C.M., Jiang, Y.H., Turner, N.D., Lupton, J.R., and Chapkin, R.S. Noninvasive detection of fecal protein kinase C βII and ζ messenger RNA: putative biomarkers for colon cancer. *Carcinogenesis* 19:253–257, 1998.
76. Jiang, Y.H., Lupton, J.R., and Chapkin, R.S. Dietary fish oil blocks carcinogen-induced down-regulation of colonic protein kinase C isozymes. *Carcinogenesis* 18:351–357, 1997.
77. Jiang, Y.H., Aukema, H., Davidson, L.A., Lupton, J.R., and Chapkin, R.S. Localization of protein kinase C isozymes in rat colon. *Cell Growth Differ.* 6:1381–1386, 1995.
78. Sauma, S., Yan, Z., Ohno, S., and Friedman, E. Protein kinase C β1 and protein kinase C β2 activate p57 mitogen-activated protein kinase and block differentiation in colon carcinoma cells. *Cell Growth Differ.* 7:587–594, 1996.
79. Saxon, M.L., Zhao, X., and Black, J.P. Activation of protein kinase C isozymes is associated with post-mitotic events in intestinal epithelial cells in situ. *J. Cell Biol.* 126:747–763, 1994.
80. Yuan, Z., Utsugisawa, T., Ishiko, T., Nakada, S., Huang, Y., Kharbanda, S., Wechselbaum, R., and Kufe, D. Activation of protein kinase C δ by the c-Abl tyrosine kinase in response to ionizing radiation. *Oncogene* 16:1643–1648, 1998.
81. Keenan, C. and Kelleher, D. Protein kinase C and the cytoskeleton. *Cell Signal.* 10:225–232, 1998.
82. Wali, R.K., Frawley, B.P., Hartmann, S., Roy, H.K., Khare, S., Scaglione-Sewell, B.A., Earnest, D.L., Sitrin, M.D., Brasitus, T.A., and Bissonnette, M. Mechanism of action of chemoprotective ursodeoxycholate in the azoxymethane model of rat colonic carcinogenesis: potential roles of protein kinase C-α, -βII, and -ζ. *Cancer Res.* 55:5257–5264, 1995.
83. Ikeda, H., Suzuki, Y., Suzuki, M., Koike, M., Tamura, J., Tong, J., Nomura, M., and Itoh, G. Apoptosis is a major mode of cell death caused by ischaemia and ischaemia/reperfusion injury to the rat intestinal epithelium. *Gut* 42:530–537, 1998.
84. Licato, L.L., Keku, T.O., Wurzelmann, J.I., Murray, S.C., Woosley, J.T., Sandler, R.S., and Brenner, D.A. In vivo activation of mitogen-activated protein kinases in rat intestinal neoplasia. *Gastroenterol.* 113:1589–1598, 1997.
85. Dove, W.F., Clipson, L., Gould, K.A., Luongo, C., Marshall, D.J., Moser, A.R., Newton, M.A., and Jacoby, R.F. Intestinal neoplasia in the ApcMin mouse: independence from the microbial and natural killer (beige locus) status. *Cancer Res.* 57:812–814, 1997.
86. Okayasu, I., Ohkusa, T., Kajiura, K., Kanno, J., and Sakamoto, S. Promotion of colorectal neoplasia in experimental murine ulcerative colitis. *Gut* 39:87–92, 1996.
87. Oshima, M., Oshima, H., Kitagawa, K., Kobayashi, M., Itakura, C., and Taketo, M. Loss of *Apc* heterozygosity and abnormal tissue building in nascent intestinal polyps in mice carrying a truncated *Apc* gene. *Proc. Natl. Acad. Sci. USA* 92:4482–4486, 1995.
88. Fazeli, A., Dickinson, S.L., Hermiston, M.L., Tighe, R.V., Steen, R.G., Small, C.G., Stoeckli, E.T., Keino-Masu, K., Masu, M., Rayburn, H., Simons, J., Bronson, R.T., Gordon, J.I., Tessier-Lavigne, M., and Weinberg, R.A. Phenotype of mice lacking functional Deleted in colorectal cancer (Dcc) gene. *Nature* 386:796–804, 1997.
89. Coopersmith, C.M., Chandrasekaran, C., McNevin, M.S., and Gordon, J.I. Bi-transgenic mice reveal that K-ras^{Val12} augments a p53-independent apoptosis when small intestinal villus enterocytes reenter the cell cycle. *J. Cell Biol.* 138:167–179, 1997.
90. Reitmair, A.H., Cai, J., Bjerknes, M., Redston, M., Cheng, H., Pind, M., Hay, K., Mitri, A., Bapat, B., Mak T., and Gallinger, S. MSH2 deficiency contributes to accelerated APC-mediated intestinal tumorigenesis. *Cancer Res.* 56:2922–2926, 1996.
91. De Wind, N., Dekker, M., van Rossum, A., van der Valk, M., and te Riele, H. Mouse models for hereditary nonpolyposis colorectal cancer. *Cancer Res.* 58:248–255, 1998.
92. Takaku, K., Oshima, M., Mioyshi, H., Matsui, M., Seldin, M.F., and Taketo, M.M. Intestinal tumorigenesis in compound mutant mice of both Dpc4 (Smad4) and Apc genes. *Cell* 92:645–656, 1998.
93. Zaidi, N.H., Pretlow, T., O'Riordan, M.A., Dumenco, L.L., Allay, E., and Gerson, S.L. Transgenic expression of human *MGMT* protects against azoxymethane-induced aberrant crypt foci and G to A mutations in the K-*ras* oncogene of mouse colon. *Carcinogenesis* 16:451–456, 1995.

94. Fazeli, A., Steen, R.G., Dickinson, S.L., Bautista, D., Dietrich, W.F., Bronson, R.T., Bresalier, R.S., Lander, E.S., Costa, J., and Weinberg, R.A. Effects of p53 mutations on apoptosis in mouse intestinal and human colonic adenomas. *Proc. Natl. Acad. Sci. USA* 94:10199–10204, 1997.
95. Deng, C., Zhang, P., Harper, J.W., Elledge, S.J., and Leder, P. Mice lacking p21$^{CIP1/WAF1}$ undergo normal development, but are defective in G1 checkpoint control. *Cell* 82:675–684, 1995.

13

NONSTEROIDAL ANTI-INFLAMMATORY DRUGS (NSAIDs), CYCLOOXYGENASES, AND THE CELL CYCLE

Their Interactions in Colon Cancer

Basil Rigas and Steven J. Shiff

The Rockefeller University
1230 York Avenue, Box 330
New York, New York 10021-6399

1. INTRODUCTION

Elucidating the pathogenesis of colon cancer, one of the commonest fatal malignancies in the Western world, represents an important challenge for biomedical science. The seminal observation that nonsteroidal anti-inflammatory drugs (NSAIDs) reduce the incidence of and mortality from colorectal cancer[1] has ushered in a new approach to the study of colon cancer.

The best-known pharmacological target of NSAIDs is the enzyme cyclooxygenase (COX), properly called prostaglandin H synthase.[2,3] COX catalyzes the formation of several prostaglandins (PGs) and thromboxane from arachidonic acid and other fatty acids. There are two isoforms of COX, COX-1, and COX-2. The former is expressed in almost all cells, whereas the latter generally is induced in response to certain stimuli, including growth factors and mitogens. COX-2, up regulated in some colon adenomas and most colon carcinomas, appears to play an important role in colon carcinogenesis.[4,5] It is generally assumed that most, if not all, of the pharmacological effects of NSAIDs can be explained through their inhibitory effect on the COX enzymes. There is, however, significant evidence that COX-independent effects may also be operative.[6]

Cancer is considered to be a cell cycle disease.[7] The fundamental biological role of the cell cycle combined with its relationship to cancer have prompted an intense

Colon Cancer Prevention, edited under the auspices of AICR
Kluwer Academic / Plenum Publishers, New York, 1999.

study of the molecular events that control the transitions through the various phases of the cell cycle.

Below, we provide an overview of the interactions between what appear to be three important factors in colon cancer, NSAIDs, cyclooxygenases, and the cell cycle. To illustrate important concepts, we draw mostly on our own work; other work is mentioned as needed, but this paper is not intended to be a thorough review of the relevant literature.

1.1. NSAIDs Affect the Cell Cycle

Early evidence suggested that NSAIDs affect the distribution of cells in the cell cycle. This work demonstrated that treatment with NSAIDs, such as indomethacin, arrests nonintestinal cells in the G_0/G_1 phase of the cell cycle.[8,9] Prompted by the accumulating evidence that NSAIDs prevent colon cancer, we reported in 1994 our observation that sulindac, and its active metabolite sulindac sulfide, profoundly affect the distribution of colon cancer cell lines in the cell cycle.[10] For example, following treatment of HT-29 cells with sulindac sulfide 150–200 μM, most of these cells remained either in G_1 phase or exited the cell cycle (G_0) (about 70–80% in G_0/G_1 vs. about 50% of controls).[11]

Other NSAIDs have similar effects.[12] For example, indomethacin, piroxicam, aspirin and naproxen, representing four important structural classes of this large family of compounds, showed a similar effect: cells were predominantly arrested in G_0/G_1 phase. Of interest, the oxidative derivative of sulindac, sulindac sulfone, which is devoid of an effect on COX, also manifested a similar cell cycle effect when confluent HT-29 cells were fed fresh medium.[13]

Common features of these effects are their concentration- and time-dependence as well as their reversibility. The relative potency of these compounds varied.[12] Although cell culture data cannot be extrapolated into intact organisms, in some cases at least the cell cycle effect of NSAIDs was observed at concentrations achievable *in vivo*.

1.2. Overview of Cell Cycle Regulation

Figure 1 provides a schematic and simplified overview of the cell cycle of higher eukaryotes. Progression through the cell cycle is controlled, in part, through dynamic interactions between cyclin-dependent kinases (cdks) and their regulatory subunits, the cyclins.[14,15] Once expressed, cyclins bind to their cdk partners in discrete phases of the cell cycle, thus contributing to the coordination of the cell cycle phases. In continuously proliferating cells, progression through the cell cycle is controlled, in part, by the kinase activity of cyclin-cdk complexes. For example: a) cdk4 or cdk6, with the D-type cyclins, control progression through G_1;[15,16] b) cdk2, with cyclin E, regulates the G_1/S transition, also thought to be the restriction point where a cell makes an irreversible commitment to proceed with one round of cell division; and c) cdc2, with the A- and B-type cyclins, the "G_2 or mitotic cyclins", controls the G_2/M transition.[14]

Various factors stimulate (e.g., epidermal growth factor or platelet derived growth factor) or inhibit (e.g., transforming growth factor-β) cell division through molecular mechanisms that ultimately affect cdk catalytic activity. In this way, the cell cycle molecular machinery is sensitive to exogenous growth signals. The products of two major gene families bind directly to and inhibit cdk complexes in response to negative regu-

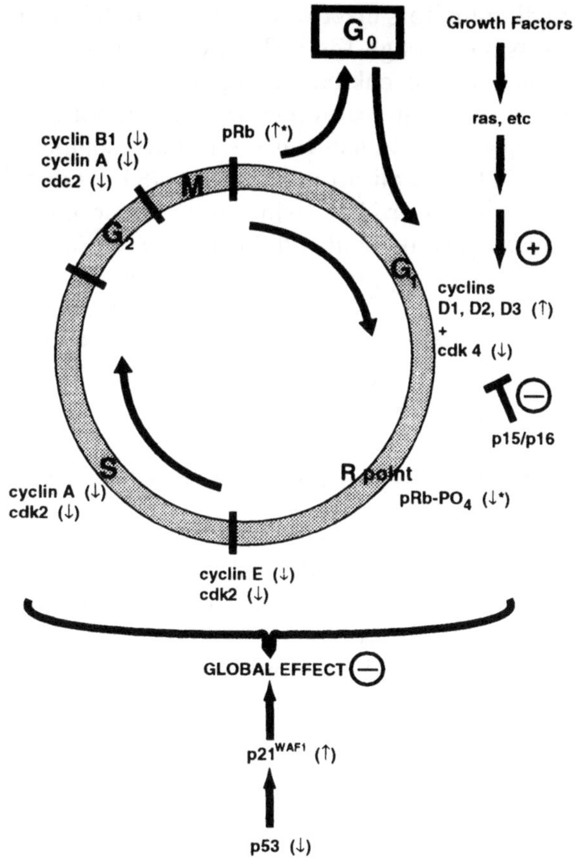

Figure 1. The cell cycle and the effect of NSAIDs. A simplified view of cell cycle progression which is controlled by dynamic interactions between cyclin-dependent kinases (cdks) and their cyclin regulatory subunits. Once expressed, cyclins bind to their cdk partners in discrete phases of the cell cycle as depicted here, ensuring the orderly advancement through the phases of the cell cycle. In continuously proliferating cells: a) cdk4 or cdk6, with the D-type cyclins, control progression through G_1; b) cdk2, with cyclin E, regulate the G_1/S transition, where the cell irreversibly commits to proceed with one round of cell division; and c) cdc2, with the A- and B-type cyclins, the "G_2 or mitotic cyclins", control the G_2/M transition. These kinases transfer inorganic phosphate to targets that affect important events in the cell cycle. For example, cdc2-cyclin B1, also known as maturation promotion factor (MPF), phosphorylates molecules like nuclear lamins, which induces the dissolution of the nuclear membrane, a dramatic event characterizing the onset of mitosis. Ubiquitin-mediated proteolysis, triggered at characteristic times in the cell cycle, degrades cyclin proteins and turns off cdk activity. The fundamental cell cycle engine is influenced by many extracellular growth factors via signal transduction pathways. For example, mitogenic factors like EGF or PDGF stimulate ras-dependent pathways which induce cyclin D expression. The tumor suppressor gene p53 inhibits progression through the cell cycle by upregulating $p21^{WAF1}$ which can inhibit the activity of many cdk complexes ("global effect"). Sulindac sulfide and other NSAIDs alter the expression of cyclins, cdks, and other molecules inhibiting cell proliferation. This effect is depicted by the direction of arrows in parentheses next to the affected protein. These extensive changes account for the induction of cell quiescence.
↑ = absolute decrease in level of protein but increase in proportion relative to the total amount; ↓* = absolute decrease in level of protein and decrease in proportion relative to the total amount. R point = restriction point; pRb = retinoblastoma protein.*

latory signals. The p15 cdk-inhibitors inactivate cyclin D-containing complexes and thus can regulate the progression through the cell cycle in early G_1. The $p21^{WAF1}$ protein family can inhibit the function of all cdk-containing complexes. Tumor suppressor genes also inhibit progression through the cell cycle. The p53 tumor suppressor gene, commonly mutated and inactivated in human cancers, is a key component of a checkpoint that regulates the onset of DNA replication near the G_1/S boundary.[17–20] p53 stimulates the expression of the cdk-inhibitor $p21^{WAF1}$ which is an important downstream effector of this checkpoint pathway. p53 likely also influences the G_2/M transition as well.[21,22] pRB, the retinoblastoma tumor suppressor protein, influences cell cycle progression in late G_1 by regulating the activity of the E2F family of transcription factors, which control transcription of a variety of genes critical for DNA synthesis.[23] The genes referred to above all play an important role in the regulation of cell proliferation and apoptosis, processes affected by NSAIDs

1.3. NSAIDs Alter the Expression and/or Activity of Cyclins and Cdks

To ascertain the mechanism by which NSAIDs produce their profound changes in the distribution of colon cancer cells in the cell cycle, we assessed their effect on the aforementioned proteins that regulate cell cycle phase transitions. At the time these studies were conducted, inhibition of colorectal tumorigenesis by sulindac was the best-documented case of NSAID-induced chemoprevention.[24,25] In fact, sulindac was the only NSAID shown convincingly to regress colonic polyps. Thus, sulindac sulfide, the active metabolite of sulindac, was the compound that we chose to study most extensively in this regard. Recently, sulindac has also been reported to decrease or even eliminate the number of aberrant crypt foci in humans.[26]

Sulindac sulfide decreases the expression of cdc2, cdk2, and cdk4 in HT-29 colon cancer cells, documented by Western blotting of proteins extracted from sulindac sulfide treated cells. Such reduction was accompanied by diminished enzymatic activity of these cdks.[11]

Sulindac sulfide profoundly affects the expression of several cyclins.[27,28] Detailed assessment of the expression of cyclins during the cell cycle phases was made possible by employing bivariate flow cytometric analysis, a methodology that allowed the simultaneous determination of both DNA and cyclin protein content of individual cells. Thus, in HT-29 cells, sulindac sulfide: a) decreases the expression of cyclins B1 and E; and b) increases the expression of cyclins D1, D2, and D3. Of interest, apoptotic cells, induced by sulindac sulfide, express both E and D-type cyclins, but not cyclin B1. In another series of experiments, it was also shown that sulindac sulfide decreases the expression of cyclin A as well.[27]

Treatment of HT-29 colon cancer cells with sulindac sulfide is associated with an increase in $p21^{WAF\text{-}1/cip1}$ levels, a reduction of the levels of pRB with a relative increase in the amount of hypophosphorylated pRB, and a reduction of the levels of mutant p53.[27] It is of interest to note that sulindac, the parent compound of sulindac sulfide, effected similar changes as sulindac sulfide, but it was less potent.[11]

These observations, summarized in Table 1 and Figure 1, make it clear that sulindac sulfide has a pronounced and fairly extensive effect on the expression and activity of molecules that regulate the cell cycle of cultured colon cancer cells (summarized in Fig. 1). Taken together, these effects are consistent with the induction of cell quiescence by sulindac sulfide and explain in molecular terms its anti-proliferative effect.

Table 1. The effect of sulindac sulfide on cell cycle of HT-29 colon cancer cells

Cell cycle phases		G0/G1 ↑↑		S ↓↓	G2/M ↓↓
Cyclins		Cdks		Tumor suppressor genes	
A	↓	cdk4	↓	pRB	↓
B1	↓	cdk2	↓	p21	↑
D1	↑	cdc2	↓	mp53*	↓
D2	↑				
D3	↑				
E	↓				

*The p53 gene is mutant in HT-29 cells.[40]

Of particular interest were two observations we made regarding the effect of sulindac sulfide on HT-29 cells. First, sulindac sulfide's inhibitory effect on cell proliferation (and by extension on cell cycle) was twice that on the induction of apoptosis.[29,30] Second, applying a triparameter flow cytometric analysis that simultaneously determined DNA content, expression of Ki-67 or PCNA, and extent of DNA strand breaks by TUNEL, we demonstrated that sulindac sulfide induced several subpopulations of cells defined by their expression of proliferation markers and DNA strand breaks. By 72h the rapidly proliferating cells [PCNA/Ki-67(+)/TUNEL(−)] were reduced from >90% to about one third. Of the remaining cells, about one third were apoptotic [PCNA/Ki-67(−)/TUNEL(+)] and one third were quiescent [PCNA/Ki-67(−)/TUNEL(−)]. Another subpopulation was detected that was PCNA/Ki-67(+)/TUNEL(+); of them, some had a dominant subdiploid peak and over half were in the S or G_2/M phase by DNA content. That a subpopulation of apoptotic cells strongly expressed PCNA and Ki-67, suggested that the specificity of these antigens as proliferation markers may need reassessment.

Although comprehensive information on the molecular effects of other NSAIDs on cell cycle is not available, they induce a similar array of changes.[12] All appear to produce the same final changes in the distribution of colonocytes in the cell cycle phases, the most notable being a predominance of cells in G_0/G_1.[12]

A major question that remains is whether NSAIDs influence the cell cycle or tumor suppressor pathways directly or through interactions with other signaling pathways at the cell membrane or within the cytoplasm. The global influence of these compounds on the molecules controlling cell proliferation and the necessity for at least 12 hrs to transpire for these effects to become manifest, suggest that they trigger unidentified upstream signaling pathways. Herrmann et al. have provided convincing evidence that sulindac sulfide binds to and inhibits the signaling function of ras,[31] a gene centrally important to mitogenic signaling and cell transformation.[32] Clearly, understanding the pathways involved in NSAID mediated chemoprevention would be of great interest for the development of new chemotherapeutic or chemopreventive agents.

The observations presented above were made in cultured colonocytes and their immediate applicability to human colon cancer is by necessity conditional. However, it is important to note that there have been several reports of altered expression of various cyclins in association with colon cancer. In one study,[33] the expression of cyclin E was elevated in 92% of colon cancers, while cyclins A and D1 were suppressed in 63% and 41%, respectively. Cyclin E gene amplification has been found in about 10% of colon cancers.[34] Finally, a histochemical study[35] showed increased expression of cyclin

D1 in about one third of colon adenomas and carcinomas. These observations underscore the notion that changes in cell cycle regulators may participate in the process of human colon carcinogenesis.

2. THE ROLE OF CYCLOOXYGENASES

A limited amount of work suggests that the expression of COX isoenzymes influences cell cycle regulation. For example, rat intestinal epithelial cells stably overexpressing the COX-2 gene, had a G_1 phase duration two to three times longer than control transfectants.[36] This change was associated with reduced cyclin D1and cdk4 levels, slightly elevated $p21^{WAF1}$ levels, and no change in the levels of cyclin D3. The authors speculated that the delay in G_1 transit may relate to the resistance of these cells to undergo apoptosis, which could affect their tumorigenic potential.

Strong evidence indicates that inhibition of COX enzymes by NSAIDs is not required for their effects on cell cycle. Indeed, this is also true of the antiproliferative and proapoptotic effects of these compounds.[29,30,37] The initial (and direct) experiment that substantiated this notion involved the study of two human colon cancer cell lines, one lacking both COX isoenzymes (HCT-15 cells), the putative target of NSAIDs, and the other expressing both COX isoforms (HT-29 cells). HCT-15 cells do not produce PGs, even when A23187, arachidonic acid or mellitin was exogenously applied to stimulate their synthesis. Alternatively, HT-29 cells produce PGE_2, $PGF_{2\alpha}$, and PGI_2, which are induced 3- to 5-fold in response to exogenous stimulation.[37]

Treatment of HCT-15 cells with sulindac sulfide or piroxicam inhibited their proliferation, altered their cell cycle phase distribution by inducing quiescence, and induced their apoptosis. Addition of exogenous PGs along with the NSAIDs failed to reverse their effect on these three parameters of cell growth. The same response was noted with HT-29 cells. Thus, it was clear that NSAIDs could affect the cell cycle machinery independently of their ability to inhibit PG synthesis.

Sulindac sulfone and salicylic acid, the hydrolytic product of aspirin, both of which are poor inhibitors of COX enzymes, also affect the cell cycle of cultured colonocytes.[13,38] These observations further support the notion that inhibition of COX by NSAIDs is not required for their effect on the cell cycle.

As we have discussed elsewhere,[39] it is conceivable that both COX-dependent and COX-independent mechanisms may be operative in bringing about the effects of NSAIDs in colon cancer, including their effect on the cell cycle. For example, if these two mechanisms affect different control pathways, and the inhibition of each one of them leads to the same end-result, then the apparently conflicting data can be reconciled. In a system like the cell cycle, where NSAIDs cause such extensive and apparently redundant changes on regulatory molecules, it would not be difficult to envision such a duality of action.

3. CONCLUSIONS

The pathogenesis of colon cancer is a complex, multistep process. NSAIDs, which profoundly affect its development, have extensive effects on the molecular regulation of the cell cycle, culminating in the induction of cell quiescence. While COX enzymes may influence cell cycle regulation, it is also apparent that this effect can be brought about by mechanisms that do not require inhibition of COX enzymes.

Biologically, the most critical question concerns the relevance of these changes to human colon cancer development. An important corollary to this would be whether alterations of the cell cycle are needed for the chemopreventive or, equally importantly, the tumor regressing effect of NSAIDs in colon cancer. At the present time, it is uncertain whether one can conclude that, since cell cycle is central to cancer, and since NSAIDs affect colon cancer so profoundly, then these changes must be pharmacologically important. Indeed, the complexity of the system suggests that a direct answer to so crucial a problem may not be within our immediate reach. Nevertheless, the stage has been set and additional work is likely to enhance our understanding in this most important area.

ACKNOWLEDGMENTS

Arthur and Rochelle Belfer Foundation; CA73298; CA29502; and AICR 96A078.

REFERENCES

1. Thun, M.J., Namboodiri, M.M., and Heath C., Jr. Aspirin use and reduced risk of fatal colon cancer. *New Engl. J. Med.* 325:1593–1596, 1991.
2. Smith, W.L., Garavito, R.M., and DeWitt, D.L. Prostaglandin endoperoxide H synthases (cyclooxygenases)-1 and -2. *J. Biol. Chemi.* 271:33157–33160, 1996.
3. Levy, G.N. Prostaglandin H synthases, nonsteroidal anti-inflammatory drugs, and colon cancer. *FASEB J.* 11:234–247, 1997.
4. Kutchera, W., Jones, D.A., Matsunami, N., et al. Prostaglandin H synthase 2 is expressed abnormally in human colon cancer: evidence for a transcriptional effect. *Proc. Natl. Acad. Sci. USA* 93:4816–4820, 1996.
5. Eberhart, C.E., Coffey, R.J., Radhika, A., Giardiello, F.M., Ferrenbach, S., and Dubois, R.N. Up-regulation of cyclooxygenase 2 gene expression in human colorectal adenomas and carcinomas. *Gastroenterol.* 107:1183–1188, 1994.
6. Abramson, S.B. and Weissmann, G. The mechanisms of action of nonsteroidal antiinflammatory drugs. *Arthritis Rheum.* 32:1–9, 1989.
7. Pines, J. Cyclins, CDKs, and cancer. *Semin. Cancer Biol.* 6:63–72, 1995.
8. Bayer, B., Kruth, H., Vaughn, M., and Beaven, M. Arrest of cultured cells in the G_1 phase of the cell cycle by indomethacin. *J. Pharmacol. Exp. Therap.* 210:106–111, 1979.
9. Bayer, G. and Beaven, M. Evidence that indomethacin reversibly inhibits cell growth in the G_1 phase of the cell cycle. *Biochem. Pharmacol.* 28:441–443, 1979.
10. Shiff, S.J., Tsai, L., Qiao, L., and Rigas, B. The effect of sulindac on cell proliferation and the cell cycle distribution of human colon cancer cells. *Gastroenterol.* 106:A440, 1994.
11. Shiff, S.J., Qiao, L., Tsai, L.-L., and Rigas, B. Sulindac sulfide, an aspirin-like compound, inhibits proliferation, causes cell cycle quiescence, and induces apoptosis in HT-29 colon adenocarcinoma cells. *J. Clin. Invest.* 96:491–503, 1995.
12. Shiff, S.J., Koutsos, M.I., Qiao, L., and Rigas, B. Nonsteroidal antiinflammatory drugs inhibit the proliferation of colon adenocarcinoma cells: effects on cell cycle and apoptosis. *Exp. Cell Res.* 222:179–188, 1996.
13. Piazza, G.A., Rahm, A.L.K., Krutzsch, M., et al. Antineoplastic drugs sulindac sulfide and sulfone inhibit cell growth by inducing apoptosis. *Cancer Res.* 55:3110–3116, 1995.
14. Morgan, D.O. Principles of CDK regulation. *Nature* 374:131–134, 1995.
15. Sherr, C.J. Mammalian G_1 cyclins. *Cell* 73:1059–1065, 1993.
16. Sherr, C.J. G1 phase progression: cycling on cue. *Cell* 79:551–555, 1994.
17. Maity, A., McKenna, W.G., and Muschel, R.J. The molecular basis for cell cycle delays following ionizing radiation: a review. *Radiother. Oncol.* 31:1–13, 1994.

18. Kastan, M.B., Onyekwere, O., Sidransky, D., Vogelstein, B., and Craig, RW. Participation of p53 protein in the cellular response to DNA damage. *Cancer Res.* 51:6304–6311, 1991.
19. Shaw, P., Bovey, R., Tardy, S., Sahli, R., Sordat, B., and Costa, J. Induction of apoptosis by wild-type p53 in a human colon tumor-derived cell line. *Proc. Natl. Acad. Sci. USA* 89:4495–4499, 1992.
20. Fritsche, M., Haessler, C., and Brandner, G. Induction of nuclear accumulation of the tumor-suppressor protein p53 by DNA-damaging agents. *Oncogene* 8:307–318, 1993.
21. Leach, S.D., Scatena, C.D., Keefer, C.J., et al. Negative regulation of Wee1 expression and Cdc2 phosphorylation during p53-mediated growth arrest and apoptosis. *Cancer Res.* 58:3231–3236, 1998.
22. Winters, Z.E., Ongkeko, W.M., Harris, A.L., and Norbury, C.J. p53 regulates Cdc2 independently of inhibitory phosphorylation to reinforce radiation-induced G2 arrest in human cells. *Oncogene* 17:673–684, 1998.
23. Weinberg, R.A. The retinoblastoma protein and cell cycle control. *Cell* 81:323–330, 1995.
24. Waddell, W.R. and Loughry, R.W. Sulindac for polyposis of the colon. *J. Surg. Oncol.* 24:83–87, 1983.
25. Giardiello, F.M., Hamilton, S.R., Krush, A.J., et al. Treatment of colonic and rectal adenomas with sulindac in familial adenomatous polyposis. *New Engl. J. Med.* 328:1313–1316, 1993.
26. Takayama, T., Katsuki, S., Takahashi, Y., et al. Aberrant crypt foci of the colon as precursors of adenoma and cancer. *New Engl. J. Med.* 339:1277–1284, 1998.
27. Goldberg, Y., Nassif, I.I., Pittas, A., et al. The anti-proliferative effect of sulindac and sulindac sulfide on HT-29 colon cancer cells: Alterations in tumor suppressor and cell cycle-regulatory proteins. *Oncogene* 12:893–901, 1996.
28. Qiao, L., Shiff, S.J., and Rigas, B. Sulindac sulfide alters the expression of cyclin proteins in HT-29 colon adenocarcinoma cells. *Int. J. Cancer* 76:99–104, 1998.
29. Qiao, L., Shiff, S.J., and Rigas, B. Sulindac sulfide induces several subpopulations of colon cancer cells, defined by PCNA/Ki-67 and DNA strand breaks. *Biochim. Biophys. Acta* 1359:222–232, 1997.
30. Qiao, L., Shiff, S.J., and Rigas, B. Sulindac sulfide inhibits the proliferation of colon cancer cells—diminished expression of the proliferation markers PCNA and Ki-67. *Cancer Lett.* 115:229–234, 1997.
31. Herrmann, C., Block, C., Geisen, C., et al. Sulindac sulfide inhibits Ras signaling. *Oncogene* 17:1769–1776, 1998.
32. Bos, J.L. Ras-like GTPases. *Biochim. Biophys. Acta* 1333:M19–31, 1997.
33. Wang, A., Yoshimi, N., Suzui, M., Yamauchi, A., Tarao, M., and Mori, H. Different expression patterns of cyclins A, D1, and E in human colorectal cancer. *J. Cancer Res. Clin. Oncol.* 122:122–126, 1996.
34. Kitahara, K., Yasui, W., Kuniyasu, H., et al. Concurrent amplification of cyclin E and cdk2 genes in colorectal carcinomas. *Int. J. Cancer* 62:25–28, 1995.
35. Arber, N., Hibshoosh, H., Moss, S.F., et al. Increased expression of cyclin D1 is an early event in multistage colorectal carcinogenesis. *Gastroenterol.* 110:669–674, 1996.
36. DuBois, R.N., Shao, J., Tsujii, M., Sheng, H., and Beauchamp, R.D. G_1 delay in cells overexpressing prostaglandin endoperoxide synthase-2. *Cancer Res.* 56:733–737, 1996.
37. Hanif, R., Pittas, A., Feng, Y., et al. Effects of nonsteroidal antiinflammatory drugs on proliferation and on induction of apoptosis in colon cancer cells by a prostaglandin-independent pathway. *Biochem. Pharmacol.* 52:237–245, 1996.
38. Elder, D.J.E., Hague, A., Hicks, D.J., and Paraskeva, C. Differential growth inhibition by the aspirin metabolite salicylate in human colorectal tumor cell lines: enhanced apoptosis in carcinoma and *in vitro*-transformed adenoma relative to adenoma cell lines. *Cancer Res.* 56:2273–2276, 1996.
39. Rigas, B. and Shiff, S.J. Is inhibition of cyclooxygenase required for the chemopreventive effect of NSAIDs in colon cancer? A model reconciling the current contradiction. *Med. Hypotheses* submitted.
40. Rodrigues, N.R., Rowan, A., Smith, M.E., et al. p53 mutations in colorectal cancer. *Proc. Natl. Acad. Sci. USA* 87:7555–7559, 1990.

ABSTRACTS

AMERICAN INSTITUTE FOR CANCER RESEARCH

Colon Cancer Prevention: Dietary Modulation of Cellular and Molecular Mechanisms

List of Presenters

#1 *Inhibitory Effect of Melatonin on 1, 2-Dimethylhydrazine-Induced Intestinal Carcinogenesis in Rats*

Vladimir Anisimov
Chief
Laboratory of Mechanisms of Carcinogenesis
N.N. Petrov Research Institute of Oncology
68, Leningradskaya Str.; Pesochny-2
St. Petersburg 189646
RUSSIA
Phone: 812/437-8607 Fax: 812/437-8947

#2 *Activated Fiber Sorbent "Aqualen" - A New Agent for Prevention of Colon Carcinogenesis*

Vladimir Anisimov
Chief
Laboratory of Mechanisms of Carcinogenesis
N.N. Petrov Research Institute of Oncology
68, Leningradskaya Str.; Pesochny-2
St. Petersburg 189646
RUSSIA
Phone: 812/437-8607 Fax: 812/437-8947

#3 *Prevention of Papillomavirus Initiated Cervical Cancer by the Phytochemical Indole-3-Carbinol*

Karen Auborn, Ph.D.
Long Island Jewish Medical Center
Albert Einstein College of Medicine
Department of Otolaryngology
270-05 76th Avenue
New Hyde Park, NY 11040
Phone: 718/470-7558 Fax: 718/347-2320

#4 *Fish Oil Constituent Docosahexaenoic Acid Inhibits Growth of Cells Rendered Pre-Cancerous by Papillomavirus*

Karen Auborn, Ph.D.
Long Island Jewish Medical Center
Albert Einstein College of Medicine
Department of Otolaryngology
270-05 76th Avenue
New Hyde Park, NY 11040
Phone: 718/470-7558 Fax: 718/347-2320

#5 *Reduction of the Incidence of Metachronus Adenomas of the Large Bowel by means of Selenomethionine and Antioxidant Vitamins: A Double Blind Randomized Trial*

Luigina Bonelli, M.D.
National Cancer Institute, Genova
Unit of Clinical Epidemiology and Trials
Largo R. Benzi, 10
16132 Genova
ITALY
Phone: 39/010 5737435 Fax: 39/010 354103

#6 *Establishing a rapid, three-dimensional culture method for chemosensitivity studies in colorectal cancer*

Peh Yean Cheah, Ph.D.
Senior Scientist
Singapore General Hospital
Department of Colorectal Surgery
Outram Park
Singapore 16908
SINGAPORE
Phone: 65/3213636 Fax: 65/2262009

#7 *The Expression of C-erb B-2, H-ras and p53 Oncoproteins in Colorectal Carcinomas and Precancerous Lesions*

Lei Dongmei, M.Sc., M.D.
Third Teaching Hospital
Henan Medical University
Department of Pathology
Kangfu Front Street
Zhengzhou Henan 450052
PEOPLE'S REPUBLIC OF CHINA
Phone: 0371/6972339 Fax: 86371/6960462

#8 *Pharmacodynamic Effects of Perillyl Alcohol in Humans*

Raymond Hohl, Ph.D.
Assistant Professor
University of Iowa
Department of Internal Medicine
Division of Hematology/Oncology, C32
200 Hawkins Drive
Iowa City, IA 52242-1081
Phone: 319/356-8110 Fax: 319/353-8383

#9 *Ki-Ras Codon 12 and 13 Mutations in Colon Carcinomas - Do They Have a Different Dietary Etiology?*

Ellen Kampman, Ph.D.
Division of Human Nutrition and Epidemiology
Dreijenlaan 1/Bode 154
6703 HA Wageningen
THE NETHERLANDS
Phone: 31-0-317-483867 Fax: 31-0-317-482782

#10 *Specific Integration Sites in Leukemias Induced by Friend Virus in Folate Deficient Mice*

Mark Koury, M.D.
Professor of Medicine
Vanderbilt University
Division of Hematology/Oncology
Medical Research Building #2 - Room 547
2220 Pierce Avenue
Nashville, TN 37232-6305
Phone: 615/936-1796 Fax: 615/936-1812

#11 *Stimulation of Apoptosis and Inhibition of Mitosis by Dietary Fish Oil Is Associated with Suppression of Neoplastic Induction in a Rat Model of Colon Carcinogenesis*

Peter Latham
Institute of Food Research
Norwich Research Park
Colney Lane
Norwich NR4 7UA
UNITED KINGDOM
Phone: 44/1603 255347 Fax: 44/1603 507723

#12 *Does Eicosapentaenoic Acid Affect the Balance Between Mitosis and Apoptosis in Cultured Colorectal Adenocarcinoma Cells?*

Elizabeth Lund
Institute of Food Research
Colney Lane
Norwich NR4 7UA
UNITED KINGDOM

#13 *Modulation of the Development and Growth of Colonic Aberrant Crypt Foci*

Bernadene Magnuson, Ph.D.
Assistant Professor
University of Idaho
Department of Food Science and Toxicology
Holm Research Center
Moscow, ID 83844-2201
Phone: 208/885-5961 Fax: 208/885-8937

#14 *Clinical application of intraperitoneal 5--fluorouracil infusion in postoperation colon cancer*

Qian Min
Fourth Teaching Hospital of Shenyang
Department of Surgery
PEOPLE'S REPUBLIC OF CHINA
Phone: 86/2486245034 Fax: 718/961-8727

#15 *Diet and Genetic Alterations in Colorectal Cancer*

Victor Moreno
Catalan Institute of Oncology
Cancer Epidemiology Service
Gran Via km 2
7 l'Hospitalet
08907 Barcelona
SPAIN
Phone: 34/932607812 Fax: 34/932607787

#16 *Cell Cycle Regulation and Induction of Programmed Cell Death in Ellagic Acid Treated Colon Cancer Cells*

Bhagavathi Narayanan, Ph.D.
Hollings Cancer Center
Medical University of South Carolina
171 Ashley Avenue
Charleston, SC 29425

#17 *Dietary Arachidonic Acid Eliminates the Anti-Tumor Effect of Eicosapentaenoic Acid in Minl + Mice*

Melissa Hansen Petrik
University of Tennessee
Department of Nutrition
229 Jessie Harris Building
1215 West Cumberland Avenue
Knoxville, TN 37996-1900
Phone: 423/974-5445 Fax: 423/974-3491

#18 *Induction of the transcription factor AP-1 in Cultured human colonic cells following exposure to bile acids and human fecal water*

Joseph J. Rafter
Associate Professor
Department of Medical Nutrition
Karolinska Institutet
Huddinge Hospital
Novum S-141 86 Huddinge
SWEDEN
Phone: 46/858583717 Fax: 46/87116659

#19 *An Abnormal Receptor for a Vitamin A Derivative in Promyelocytic Leukemia*

Robert Redner, M.D.
Assistant Professor of Medicine
University of Pittsburgh
Division of Hematology/Oncology
E1058 BST
200 Lothrop Street
Pittsburgh, PA 15213
Phone: 412/624-9598 Fax: 412/383-9544

#20 *Hydroxyoctadecadienoic Acid (13-HODE) is the Mitogenic Signal for Linoleic Acid-Dependent Growth in Rat Hepatoma 7288CTC in vivo*

Leonard Sauer, M.D., Ph.D.
P.O. Box 3
Stevensville, MT 59870
Phone: 406/777-4360 Fax: 406/777-4360

#21 *Modulation of Gap Junction-Connexin Gene Expression by the Chemopreventive Bowman-Birk Protease Inhibitor*

MaryJean Sawey, Ph.D.
Research Associate Professor
Temple University
Department of Radiation Oncology
3401 North Broad Street
Philadelphia, PA 19140
Phone: 215/707-8184 Fax: 215/707-8062

#22 *MTHFR Polymorphism and Risk of Colorectal Adenomatous Polyps*

Neli Ulrich, M.S.
Fred Hutchinson Cancer Research Center
1100 Fairview Avenue North, MP-702
Seattle, WA 98109-1024

#23 *P53 Overexpression vs. P53 Mutations in Relation to Dietary Risk Factors for Colon Cancer*

Dorien Voskuil, M.Sc.
Division of Human Nutrition and Epidemiology
Dreijenlaan 1/Bode 154
6703 HA Wageningen
THE NETHERLANDS
Phone: 31-0-317-483312 Fax: 31-0-317-482782

#24 *Aspirin supresses the mutator phenotype associated with HNPCC by genetic selection*

Sabine Wallinger
University of Regensburg
Institute of Pathology
Franz-Josef-Strauss-Allee 11
93053 Regensburg
GERMANY
Phone: 49/941-9446623

#25 *Dietary Arachidonic Acid Eliminates the Anti-Tumor Effect of Eicosapentaenoic Acid in Minl + Mice*

Jay Whelan
Professor
University of Tennessee
Department of Nutrition
1215 West Cumberland Avenue, Room 229
Knoxville, TN 37996-1900
Phone: 423/974-5445 Fax: 423/974-3491

Poster Abstract # 1

Activated Fiber Sorbent "Aqualen" - A New Agent For Prevention of Colon Carcinogenesis

V.N. Anisimov, M.A. Zabezhinski, I.G. Popovich, A.I. Lieberman*, J.L. Shmidt*
N.N. Petrov Research Institute of Oncology, Pesochny-2 .
*Aquaphor Corporation, St. Petersburg 189646 RUSSIA

The effect of activated carbon fiber adsorbent Aqualen on colon carcinogenesis was firstly studied in rats. Two-month-old outbred female LIO rats were weekly exposed to 15 (experiment 1, groups 1, 2 and 3) or to 5(experiment 2, groups 4, 5 and 6) subcutaneous injections of 1, 20dimethylhydrazine (DMH) at a single dose of 21 mg/kg of body weight. From the day of the first injection of the carcinogen DMH, the rats from groups 2, 3, 5 and 6 were given Aqualen with diet. In both experiments, rats were fed Aqualen 5 times per week together with lab chow in the daily dose of 10 mg/kg (groups 2 and 4) or 100 mg/kg (groups 3 and 5) of body weight. Additionally, another group of rats were not exposed to the carcinogen and served as intact control (group 7) or were given Aqualen with a diet in the daily dose of 10 mg/kg (group 8) or 100 mg/kg (group 9). These experiments were finalized in 6 months after the first injection of DMH. In experiments 1 and 2, the majority of tumors were localized in the descending colon. Tumors of the small intestines developed only in rats from experiment.

1) Total incidence of colon tumors as well as tumors in different parts of the colon and the mean number of tumors per rats were much higher in rats from all groups in experiment 1 than rats in experiment 2.
2) In experiment 1, supplementation of Aqualen to diet was followed by the decrease in the number of tumors per rat in both ascending and descending colon regardless of the dose of the enterosorbent.

In experiment 2, the affect of Aqualen was stronger than that in experiment 1: the enterosorbent decreased both the incidence and the multiplicity in the total colon, its ascending and descending parts and in rectum. In both experiments the percentage of small colon tumors among rats exposed to Aqualen (groups 2 and 3) was higher than that of controls (group 1). All detected intestinal tumors were classified as adenocarcinomas. There was not any differences in the tumor morphology between rats from groups exposed to Aqualen without DMH. Thus, our results demonstrate the inhibitory effect of activated carbon fiber adsorbent Aqualen on intestinal carcinogenesis in rats.

Poster Abstract # 2

Inhibitory Effect of Melatonin on 1, 2-Dimethylhydrazine-Induced Intestinal Carcinogenesis in Rats

V.N. Anisimov, I.G. Popovich, M.A. Zabezhinski, S.V. Musatov, A.V. Arutjunyan, I.M. Kvetnoi, I. Zuzman.
N.N. Petrov Research Institute of Oncology, Pesochny-2
St. Petersburg 189646 RUSSIA

Three-month-old male LIO rats were exposed to 5 or 15 weekly s.c. injections of 1, 2 dimethylhydrazine (DMH, 21 mg/kg) and were given Melatonin in tap water (20 ug/ml) or solvent at night time during the period of the carcinogen treatment. Rats were sacrificed 6 months after the start of DMH injections. The exposure to Melatonin (MLT) was followed by the decrease in intestinal tumor incidence in jejunum and ascending colon as compared to controls. The inhibitory effect of MLT on the multiplicity of bowel tumors was found mostly in duodenum, jejunum, ascending and descending colon. This effect is correlated with the significant inhibitory effect of the pineal hormone on mitotic index and with stimulating inhibitory effect of MLT on the relative number of apoptotic cells (TUNEL-method) in colon tumors. Long-term treatment with MLT was followed also by the decrease in the area of lymphoid infiltrates in the colon mucosa of tumor-bearing rats. The level of immunohistochemically detected MLT in epithelium of the intestinal tract of rats exposed to DMH was significantly reduced as compared to intact rats, while in rats treated with DMH+MLT it was in control ranges. In serum of rat exposed to DMH alone the level of diene conjugates (DC) and Schiff's bases (SB) was significantly increased as compared with controls. In colon tissue of DMH-treated rats, the level of DC, SB, carbonyl derivatives of amino acids NO-synthase activity was significantly increased and total antioxidative activity was decreased as compared to controls. In rats exposed to DMH+MLT normalization of free radical processes in serum and colon have been observed. MLT also inhibits mutagenic effect of DMH in vivo (chromosome aberration and sperm head anomalies tests) and in vitro (Ames test). These findings suggest that the inhibitory effect of MLT on the intestinal carcinogenesis is mediated by an inhibition of free radical processes in the target for DMH tissues.

Poster Abstract # 3

Prevention of Papillomavirus Initiated Cervical Cancer by the Phytochemical Indole-3-Carbinol

Liang Jin, Mai Qi, DaZhi Chen and Karen Auborn
Department of Otolaryngology, Long Island Jewish Medical Center, The Long Island Campus of Albert Einstein College of Medicine, New Hyde Park, NY 11040

Cervical cancer is the second most common cancer worldwide. The E6 and E7 proteins of the highly oncogenic types of papillomaviruses, such as human papillomavirus type 16, are cofactors for cervical cancer. Other factors are needed for cervical cancer to develop. That estrogen promotes the development of cervical cancer is indicated by much circumstantial and direct evidence. We asked whether indole-3-carbinol (I3C) could prevent cervical cancer in a mouse model. I3C, found in high concentrations in cruciferous vegetables (broccoli, cabbage, cauliflower and brussel sprouts), is an active chemopreventative. In particular, I3C is an anti-estrogen by favorably modulating estrogen metabolism.

Mice with the transgene for early genes of HPV16 with a keratin 14 promoter develop cervical cancer when given estradiol chronically (Arbeit et al., PNAS 93:2930, 1996). Using these mice, we compared mice fed a diet AIN or AIN supplemented with 2000 ppm of the phytochemical, indole-3-carbinol. At a dose of estradiol (0.125mg/60 days), many mice became morbid caused by retention of fluid in the bladder. This morbidity was reduced by I3C. More pre-cancerous lesions developed with time and more cancers developed with time. I3C prevented the cancers and reduced the development of precancerous lesions. We used lower doses of estradiol (0.125mg/60 days) to prevent morbidity effects associated with estradiol. More than half of these mice developed invasive cervical cancer within 6 months, and the remainder developed high grade precancerous lessons. In the mice receiving dietary indole-3-carbinol, no mice developed invasive cervical cancers and fewer developed precancerous lesions.

We conclude that I3C would be likely be useful in the prevention of cervical cancer and effective in individuals known to have papillomavirus infections.

Poster Abstract # 4

Fish Oil Constituent Docosahexaenoic Acid Inhibits Growth of Cells Rendered Pre-Cancerous by Papillomavirus:

DaZhi Chen and Karen Auborn
Department of Otolaryngology, Long Island Jewish Medical Center, The Long Island Campus of Albert Einstein College of Medicine, New Hyde Park, NY 11040

The omega -3 PUFAs (polyunsaturated fatty acids) inhibit proliferation of breast cancer cells whereas omega-6 PUFAs stimulate growth. In this study, we examined papillomavirus infected cells. Both genital cells immortalized with the highly oncogenic human papillomavirus type 16 (HPV16) and cells grown from explants of benign tumors caused by papillomavirus type 6 or 11 were examined.

Human cervical keratinocytes, immortalized with HPV16, were treated with linoleic acid (an omega-6 PUFA) and the omega-3 PUFAs, eicosapentaenoic acid (EPA) and docosahexaenoic acid (DHA). Using both cell counts and BrdU incorporation, DHA inhibited growth to a greater extent than EPA, whereas linoleic acid had no effect. This effect of DHA was dose-dependent, and largely growth arrest rather than toxicity. DHA inhibited growth of foreskin keratinocytes immortalized by HPV16 and cells grown from the explants of laryngeal papillomas, a benign tumor with an HPV etiology, but did not inhibit normal foreskin or laryngeal keratinocytes. The effect of indomethacin, a cyclooxygenase inhibitor like DHA, had only minimal effect on growth. DHA inhibited growth in the presence of estradiol, a grwoth stimulator. Alpha-tocopherol, a peroxidation inhibitor, largely abrogated effects of DHA. Therefore, the mechanism of inhibition is consistent with lipid peroxidation.

Poster Abstract # 5

Reduction of the incidence of metachronous adenomas of the large bowel by means of selenomethionine and antioxidant vitamins: A double blind randomized trial.

Luigina Bonelli*, Annalisa Camoriano', Paolo Ravelli#, Guido Missale#, Paolo Bruzzi* and Hugo Aste'§.
* Unit of Clinical Epidemiology and Trial, and ' Dept. of Gastroenterology and Digestive Endoscopy; National Cancer Institute, Genova; Dept of Medical Oncology University of Genova Italy. # Dept. of Surgical Endoscopy, University of Brescia, Italy

Purpose:: to investigate the efficacy of a combination of antioxidants in reducing the incidence of colorectal metachronous adenomas after polypectomy.

Methods: Between 1988 and 1995, in two endoscopic units, 304 patients who had at least one adenoma removed from the large bowel and were free from polyps were randomized to receive daily, for 5 years, 200 mcg. l-selenomethionine, 30 mg. zinc, 2 mg vitamin A, 180 mg. vitamin C and 30 mg. vitamin E, or a placebo (both provided by Pharma Nord, Vejle, Denmark). Patients were scheduled for colonoscopy on year one, three and five after randomization. An intention to treat analysis was performed.

Results: 233 of 304 patients had at least one follow-up colonoscopy (116 in the intervention group and 117 in the placebo group). Their 5-year actuarial compliance to the treatment was 63.7%; the results were different in the two centers (40.5% vs 96.1%). The four-year adenoma recurrence-free survival was 76.9% in the intervention group and 64.2% in the placebo group (P=0.035). The observed incidence of metachronous adenomas was 5.9% in the intervention group and 11% in the placebo group (crudeRR=0.51, c.i. 0.27-0.95, adjRR=0.56, c.i. 0.32-0.99); the reduction of the risk was more marked in the center were better compliance was obtained (RR=0.23, c.i. 0.06-0.92) than in the other (RR=0.64, c.i. 0.30-1.35). Side effects related to the treatment were reported by 12.5% of the patients in the intervention group and by 2.9% of those in the placebo group.

Conclusions: Despite its limitations (low compliance, high rate of lost to follow up), this study provides suggestive evidence in favor of the preventive effect of the chemopreventive regimen on the incidence of metachronous adenomas. Furthermore, in the center where compliance to treatment and follow up was more than adequate (>90%), the observed effect was even more marked.

Poster Abstract # 6

Establishing a rapid, three-dimensional culture method for chemosensitivity studies in colorectal cancer. Cheah PY., Liu, Y, Eu KW, Seow C. Dept. of Colorectal Surgery, Singapore General Hospital, Singapore 169608

Traditionally, cell lines have been used for drug studies. However, monolayer cell cultures are much more sensitive to drugs than the in vivo situation. On the other hand, nude mice xenograph takes 30-50 weeks to yield results and is therefore of limited use to most cancer patients. Moreover, results obtained in xenographs may not be representative of human response.

We recently established a three-dimensional histoculture system that is able to simulate the in vivo microenvironment for chemopreventive studies in colorectal cancer (CRC). Fresh CRC specimens were dissected and cultured on pig skin collagen floating on tissue culture medium and grew in CO_2 incubator at 37°C. The viability of the tumor blocks were evaluated by the BrdU immunostaining method. The average labeling index is ~50%. The histoculture tumors were able to maintain the typical tissue architecture of CRC tumors and the expression of tumor specific antigen after culture for 14 days.

The histoculture system has clear advantages over cell line and xenograph studies as it is more in vivo-like than cell lines and more rapid and cost-effective than xenographs. Moreover, it offers potential for individualized treatment since each culture involves tumor tissues from individual patient. It is also a good system for tumor biology studies and for screening new anti-cancer drugs.

Preliminary results on the chemopreventive effect of staurosporine, a potent Protein Kinase C inhibitor, as measured by the proliferative and apoptotic indices, will be presented.

Poster Abstract # 7

The Expression of C-erbB-2, H-ras and p53 Oncoproteins in Colorectal Carcinomas and Precanerous Lesions

Lei Dongmei [1)] Yin Zhirong[2)] Hao Zhifang[3)]

1) Department of Pathology, Third Teaching Hospital, Henan Medical University, Zhengzhou, 450052, P.R.China
2) Department of Pathology, First Teaching Hospital, Henan Medical University, Zhengzhou, 450052, P.R.China

Abstract

Objective: To investigate the relationship of the expression of C-erbB-2, H-ras, p53 gene proteins and development of colorectal carcinoma and its clinical value in the early diagnosis, and prognosis in colorectal carcinoma. **Methods**: With LSAB immunohistochemistry, the paraffin embedded tissues of 45 cases of colorectal carcinoma and 36 cases of non-adenomas dysplasia abjecent to colorectal carcinoma were studied. **Results:** p53 positive expression rate in colorectal carcinoma is 57.8%, non-adenomas dysplasia abjecent to colorectal carcinoma and colorectal adenoma which show high expression are both moderately or high dysplasia. The overexpression of p53 relates with histologic grade, lymph node metastasis (P<0.05), but no significant correlation the expression of p53 and histologic type, degree of invasion. Both of p21 and p185 expression does not correlate with histologic type, degree of invasion lymph node metastasis, but histologic grade (<0.05). 29(64.4%) cases showed coordinately positive expression for p21 and p185(P<0.005). **Conclusion:** p53 expression may be an independent marker of malignant potential. The positive expression of p21 correlates with that of p185 strongly p53, p21 and p185 play an important role in the development of colorectal carcinomas.

Key word: Colorectal Carcinoma C-erbB-2 ras p53

Poster Abstract # 8

PHARMACODYNAMIC EFFECTS OF PERILLYL ALCOHOL IN HUMANS.

R.J. Hohl, K. Lewis, G. Ripple, M. Gould, R. Marnocha, R. Arzoomanian, D. Alberti, K. Tutsch, C. Feierabend, K. Simon, K. Morgan, J. Stewart, A. Wahamaki, M. Pomplum, G. Wilding, and H. Bailey. University of Iowa College of Medicine, Iowa City, IA and University of Wisconsin, Madison, WI.

Perillyl alcohol is a naturally occurring monoterpene which has shown antitumor and cancer preventative activity in preclinical studies. Because of these properties and that it is a dietary constituent there is intensive interest in defining the underlying mechanism(s) for its anticancer effects. We have previously demonstrated that perillyl alcohol *in vitro* in high concentrations may inhibit farnesyl protein transferase (FPTase), which is the enzyme that catalyzes RAS post-translational processing, and in lower concentrations will reduce RAS levels. We have hypothesized that these effects may contribute to perillyl alcohol's anticancer activity. To investigate whether these *in vitro* effects are observed *in vivo* we are conducting a phase I trial of perillyl alcohol. Perillyl alcohol is given four times daily for two weeks followed by two weeks off. Ten patients have been treated at the following doses: level (L) 1-1200 mg/m2/dose; L2-1600 mg/m^2/dose. Preliminary pharmacokinetic data is available. Peak levels (μM) of the two main metabolites, perillic acid (PA) and dihydroperillic acid (DHPA), were 200-350 for PA and 20-30 for DHPA. Day 14 and 29 metabolite levels were similar to day 1. Metabolite levels appear to increase from L1 to L2 and there is a trend toward higher levels when the drug is taken in a fasting state than when taken with food. The main toxicities observed have been gastrointestinal (nausea and vomiting, satiety and gastroesophageal reflux) and fatigue. Correlative laboratory studies are also being evaluated from a mechanistic and prognostic standpoint. Plasma obtained after perillyl alcohol dosing does not display an ability to inhibit mammalian FPTase in an *in vitro* bioassay. Mononuclear cell RAS levels vary after perillyl alcohol treatment and decreased in 3 of the first 7 patients analyzed with the most dramatic decrease being to 30% of control in a patient at L2. These correlative findings are encouraging and support the possibility that alterations in RAS can occur *in vivo*. Enrollment continues to determine the maximum tolerated dose when given on this schedule. Supported by the NIH and the AICR.

Poster Abstract # 9

Ki-RAS CODON 12 AND 13 MUTATIONS IN COLON CARCINOMAS - DO THEY HAVE A DIFFERENT DIETARY ETIOLOGY?

Ellen Kampman[1], Dorien W. Voskuil[1], Annemieke A. van Kraats[2], Helena F. Balder[1], Goos N. P. van Muijen[2], R. Alexandra Goldbohm [3], Pieter van 't Veer[1]

[1]Division of Human Nutrition and Epidemiology, Wageningen Agricultural University, The Netherlands [2]Department of Pathology, Nijmegen University Hospital, Nijmegen, The Netherlands.[3] Department of Consumer Research and Epidemiology, TNO Nutrition & Food Research Institute, Zeist, The Netherlands.

The involvement of the Ki-ras oncogene and its specific mutations in colon tumors may be associated with dietary habits.

Based on a previously conducted case-control study on diet and colon cancer, we collected paraffine-embedded carcinoma tissue of the cases to study the potential association between Ki-ras mutational status and diet. Ki-ras gene mutations (codons 12 and 13) were determined by a mutant allele specific method: 36% of the 185 cases harbored mutations, of which 82% were located in codon 12. Overall, Ki-ras mutational status was not markedly associated with diet. Interestingly, however, high intakes of animal protein, calcium and poultry were differently associated with codon 12 and codon 13 mutations: odds ratios (95% confidence intervals) codon 12 versus codon 13 were 9.0 (2.0-42), 4.1 (1.4-12) and 15 (1.4-160), respectively. Differences between the codons could not be explained by the type of mutations involved: transition and transversion mutations were not differently associated with diet.

These data suggest a different dietary etiology of colon tumors harboring Ki-ras codon 12 and codon 13 mutations.

Poster Abstract # 10

Specific Integration Sites in Leukemias Induced by Friend Virus in Folate Deficient Mice, Don J. Park and Mark J. Koury, Vanderbilt University and Nashville VA Medical Center, Nashville, TN.

Folate deficiency around the time of infection with the "anemia-inducing" strain of Friend virus complex increases the incidence of leukemia over the subsequent year when compared to control mice that were not folate deficient at the time of infection. Many of these Friend virus induced leukemias can give rise to clonal leukemia cell lines. Analysis of these cell lines derived from mice that were folate-deficient at the time of infection indicate that the two most common genetic changes reported in these leukemias, *Spi-1 (Pu.1)* rearrangement and *p53* mutation, cannot account for the increased incidence of leukemia. In order to circumvent cross hybridization with endogenous murine retroviruses, we have used two dimensional Southern blotting to analyze possible integration sites by the two components of the Friend virus complex in leukemia cell lines derived from mice that were folate deficient or normal at the time of infection. The total number of provirus integrations was increased in many of the cell lines derived from folate-deficient mice as compared to those derived from controls. Furthermore, we have identified two integration sites present in some cell lines from folate deficient mice that are not present in cell lines of normal mice. We are in the process of cloning the murine DNA near these sites to determine whether they are in or near recognized folate fragile areas, oncogenes, or tumor suppressor genes. These studies should help identify possible genetic loci that are altered by the combination of folate deficiency and leukemogenic stimuli.

Poster Abstract # 11

Stimulation of apoptosis and inhibition of mitosis by dietary fish oil is associated with suppression of neoplastic induction in a rat model of colon carcinogenesis.

Peter Latham, Elizabeth K. Lund, and Ian T. Johnson
Institute of Food Research - Norwich Laboratory
Norwich Research Park
Colney, NORWICH,
NR4 7UA, UK.

Diets rich in the n-3 polyunsaturated fatty acids (n-3 PUFA) eicosapentaenoic acid (C20:5, EPA) and docosahexaenoic acid (C20:6, DHA) derived from marine fish have been shown to suppress the genesis of preneoplastic aberrant crypt foci (ACF) in animals administered carcinogen (1), the development of microadenomas in the APC (Min) knockout mouse (2), and tumour development and growth in rodents (3). The mechanisms of this protection are largely unknown. In this study we wished to gain insight into the effects of dietary fish oils on crypt cell apoptosis and mitosis during a 48 h period following administration of 1,2-dimethylhydrazine (DMH), and to investigate whether modifications to cell fate induced by fish oil may constitute a mechanism by which it can suppress the induction of ACF in the rodent distal colon. The consumption of fish oil was associated with increased apoptotic cell death and reduced proliferation in the colon at 24 h and 48 h after administration of DMH. Apoptosis was also increased, and mitosis reduced from baseline levels in control animals given a sham injection of saline and fed fish oil. After 18 weeks, animals fed fish oil for 48 h after each of three weekly DMH injections had approximately 50% fewer ACF, although there were no significant differences in the numbers of aberrant crypts per focus. These cytokinetic and apoptotic responses of the rat colonic crypt to fish oil after treatment with a model carcinogen, coupled with the observed suppression of ACF, suggest that dietary PUFAs can mediate changes in the balance of cell proliferation and death in the colonic mucosa, and hence modulate the induction of early neoplastic lesions.

References

1) Paulsen, J. E., Stamm, T., and Alexander, J. A fish oil-derived concentrate enriched in eicosapentaenoic and docosahexaenoic acid as ethyl esters inhibits the formation and growth of aberrant crypt foci in the rat colon. *Pharmacol. Toxicol,* **82(1)**: 28-33, 1998.

2) Paulsen, J. E, Elvsaas, I. K., Steffensen, I. L., and Alexander, J. A fish oil-derived concentrate enriched in eicosapentaenoic and docosahexaenoic acid as ethyl esters suppresses the formation and growth of intestinal polyps in the Min mouse. *Carcinogenesis*, **18(10)**: 1905-1910, 1997.

3) Reddy, B. S., and Maruyama, H. Effect of Dietary Fish Oil on Azoxymethane-induced Colon Carcinogenesis in Male F344 Rats. *Cancer Res.*, **46**: 3367-3370, 1986.

Poster Abstract # 12

Does eicosapentaenoic acid affect the balance between mitosis and apoptosis in cultured colorectal adenocarcinoma cells?

Rosemary G Clarke, Ian T Johnson, Elizabeth K Lund. Institute of Food Research, Colney, Norwich UK. NR4 7UA.

Studies in animal models have shown that fish oil supplements can both reduce colonic crypt cell proliferation rates in healthy animals[(1)] and reduce the numbers of aberrant crypt foci created in response to exposure to chemical carcinogen[(2)]. An increase in crypt cell apoptosis has been implicated in the reduction in the numbers of precancerous[(3)] and cancerous [(4)] lesions seen in animal models in response to dietary factors.

The aim of this study was to investigate the effects of eicosapentaenoic acid (EPA), at levels equivalent to those which are achievable in plasma, on the balance between cell division and cell death using HT29 cells as a model for colon cancer. We have studied both the adherent and floating cells as separate populations to more fully understand the process of cell growth, differentiation and death. Our results indicate that although EPA causes a reduction in total adherent cell number this is not explained by change in the rate of cell division but is due to loss of cells into the medium. Using flow cytometric analysis combined with a range of techniques including staining with propidium iodide, TUNEL assay, and annexin V we have shown that none of the classical features of apoptosis are seen in adherent HT29 cells exposed to EPA. This contrasts with indomethacin used as a positive control. All floating cell populations contained normal, apoptotic and necrotic cells but the proportions of each varied between controls, indomethacin and EPA treated cells. Our results suggest that the apoptosis induced by EPA, in the crypt in-vivo, may result from loss of cell-cell or cell-matrix contact through changes in adhesion molecules similar to that described previously for monocytes [5]. We are currently investigating these phenomena in relation to a number of chemoprotective dietary factors.

1. Pell, J. D., Brown, J. C., and Johnson, I. T. Polyunsaturated fatty acids of the n-3 series influence intestinal crypt cell production in rats. Carcinogenesis, 15: 1115-1119, 1994.

2. Latham P, Lund E.K. and Johnson I.T. Inhibition of mitosis and stimulation of apoptosis by dietary fish oil is associated with suppression of neoplasia in the rat. Proceedings of Colon Cancer Prevention: Dietary Modulation of Cellular and Molecular Mechanism. 1998

3. Smith, T. K, Lund, E. K., and Johnson, I. T. Inhibition of dimethylhydrazine-induced aberrent crypt foci and induction of apoptosis in rat colon following oral administration of the glucosinylate sinigrin. Carcinogenesis, 19: 267-273, 1998.

4. Chang, W-C., L., Chapkin, R. S., and Lupton, J. R. Predictive value of proliferation, differentiation and apoptosis as intermediate markers for colon tumorigenesis. Carcinogenisis, 18: 721-730, 1997.

5. Hughes, D. A., Pinder, A. C., Piper, Z., Johnson, I. T., and Lund, E. K. Fish oil supplementation inhibits the expression of major histocompatibility complex class II molecules and adhesion molecules on human monocytes. Am. J. Clin Nutr., 63: 267-272, 1996.

Poster Abstract # 13

Modulation of the development and growth of colonic aberrant crypt foci

B.A. Magnuson, E.H. South, S. Hubele, K. Hendrix, J.H. Exon
Department of Food Science, University of Idaho, Moscow, ID

(ACF) has been widely used as a method to identify dietary cancer. The role of aging in the development of these lesions has not been previously investigated. 38 Sprague Dawley rats were treated with 2 injections of azoxymethane (AOM) (15 mg/kg body weight one week apart) at a young (6 weeks, n=20) or older (52 weeks, n=18) age. Rats were killed 6 and 14 weeks following the last AOM injection, and the number, size and location of ACF were determined. Cell proliferation was measured immunohistochemically using incorporation of bromodeoxyuridine (BrdU) into DNA. At both the 6 and 14 week time points, the number of ACF in older rats was significantly greater ($p<10.01$) than the number of ACF in young rats. There was also a significant difference in the distribution of ACF throughout the colon. In older rats, an increased incidence of ACF was evident in the colonic region from the mid-point to the cecum as compared to the incidence in young rats. In both the old and the young rats, there was a significant increase in the number of large ACF after 14 weeks, as compared to after 6 weeks. There were a higher number of larger ACF in older animals compared to young animals at both time points, however there was no significant differences in the percentage of all ACF that were large (4-6 crypts/foci) and very large (greater than 6 crypts/foci). Differences in the indices of colonic cell proliferation between old and young rats do not appear to be sufficient to be responsible for differences in ACF development. This data demonstrates that age-associated changes in colonic cells alters the development and distribution of AOM-induced ACF.

Poster Abstract # 14

Clinical application of intraperitoneal 5—fluorouracil infusion in postoperation colon cancer

Qian Min

Department of Surgery, The Forth Hospital of Shenyang P.R.China

ABSTRACT: Thirty two patients with colon cancer followed surgical in order to investigate the clinical value of post-intraperiloneal 5—FU infusion on colon cancer. All 32 patients with advanced colon cancer received 148 times in 88 courses of intraperitoneal 5—FU infusion of high doses in volume of 5—FU. 32 cases finished 1 course, 30 cases finished 2 courses, 26 cases 3 courses. Hematologic,hepatic and renal toxicities as well as serious complications, such as intestinal perforation, bleeding, anastomotic leakage, infection and serious peritoneal adhesions in peritoneal cavity, were not observed, antitumor efficacy was better. Results showed that intraperitoneal 5FU infusion was a safe simple adjuvant treatment without serious complications for postoperative patients with colon cancers. 28 cases followed up for 2—5 years, average of 42 months. Recurrence 3 cases, among which 2 cases died of intestinal obstruction due to local implantation. I case died of pulmonary metastasis.

Key words: colon cancer, 5—fluorouracil, intraperitoneal infusion.

Poster Abstract # 15

Diet and Genetic Alterations in Colorectal Cancer

Interim Analysis of a Genetic Epidemiology Case-Control Study.

Moreno, V*, Bosch, F.X., Peinado, M., Gonzalez, I., Marti, J., Navarro, M., Cambray, Benasco, C. and the Bellvitge Colorectal Cancer Study Group.
Cancer Epidemiology Service.
Catalan Institute of Oncology.
Gran Via km 2, 7 l'Hospitalet
08907 Barcelona SPAIN

A hospital based case-control study of colorectal cancer is being conducted to assess the relationship between diet and genetic alterations found in the tumors. Cases and controls are interviewed with a structured questionnaire on risk factors and dietary history. For operated cases, samples of tumor and normal tissue are collected and frozen. After DNA extraction, k-ras oncogene is screened for mutations in codons 12 and 13 of exon 1 by PCR amplification and SSCP separation. Dubious cases are sequenced. Mutations in p53 gene and genetic instability are also being explored with similar methods. Only exploratory food group analysis has been done. Classification of cases according to genetic alterations allows different logistic regression modeling for risk estimation. Beyond classical analysis (all cases vs. Controls), a case-case comparison is done and the risk of colorectal cancer related to food groups is estimated for tumors with and without mutations. Currently 380 cases have been interviewed out of 446 identified and 362 controls out of 395. Tumor tissues of 201 cases have been analyzed for mutations in k-ras (41% of them mutated). 178 cases have been analyzed for p53 (59% mutated). High total caloric intake is a major risk factor for colorectal cancer (average of 49% increase in risk per 1000 kcal, 95% CI = 18%-87%). Analyses of associations of food groups with colorectal cancer show expected results of a protective effect of vegetables, especially those consumed raw in salads (OR = 0.86 per 100 grams daily, 95% CI = 0.76 to 0.98). Similarly to previous work (Bautista et al. CEBP 1997; 6:57-61), initial screening of differences in food group consumption between cases with mutations in k-ras or p53 and cases with wild type genes has not shown very strong associations, but detailed nutrient analysis is ongoing.

Poster Abstract # 16

Cell cycle regulation and induction of programmed cell death in Ellagic acid treated colon cancer cells

Bhagavathi A. Narayanan, Otto Geoffroy, *Gian G. Re, Thomas Gue and Daniel W. Nixon, Cancer Prevention Program, Hollings Cancer Center and *Dept. of Pathology and Laboratory Medicine, Medical University of South Carolina, Charleston, SC 29425

Colorectal carcinoma is the second leading cause of cancer mortality in the United States. A number of predisposing factors have been identified for colorectal cancer. Although a close epidemiological correlation between diet and colorectal cancer has been reported, knowledge about the cellular and molecular relationships between diet and colon cancer is presently unclear. The use of phytochemicals as chemopreventive agents is attracting interest in reducing the risk of colorectal and other cancer. Ellagic acid, a naturally occurring plant phenolic compound, is present in high concentration in nuts and berries as both a monomer and as a complex mixture of oligomeric ellagitannins. Ellagic acid has been demonstrated to inhibit cancer growth in animals induced by several chemical carcinogens including polycyclic aromatic hydrocarbons, N-nitrosoamines, aflatoxins and aromatic amines in animal models. Other studies have demonstrated that ellagic acid, applied topically on mouse skin, could effectively inhibit TPA-induced ornithine decarboxylase activity, hydroperoxide production and DNA synthesis. Oral administration of ellagic acid was found to significantly reduce the level of lipid peroxidase and liver dihyroxy proline in animal models. We previously have shown (AACR Proceedings on Molecular Mechanism of Apoptosis regulation, 1998) that exposure of cervical carcinoma cells in culture to ellagic acid, at concentrations of 10^{-5} M induced apoptosis, activated p53, and unregulated p21 $^{WAFI/CIPI}$ gene and arrested cell cycle proression in the G1 phase

In the present study, we evaluated the effect of ellagic acid on the cell cycle regulation of colon cancer cells (SW480) in culture. The effects of ellagic acid on cell cycle events and apoptosis were analysed by TdT assay and FACS analysis. We found that ellagic acid at a concentration 10^{-5} M induced cell cycle arrest at G1-S within 48 hrs, inhibited overall cell growth and induced apoptosis (32.7%) after 72 hrs of treatment, when compared to the cell which are untreated. Ellagic acid treated cells stained with 4, 6-diamidine-2'phenylindole dihydrochoride (DAPI) and subjected to DNA fragmentation analysis further demonstrated the nuclear condensation and internucleosomal DNA strand breaks characteristic of apoptosis. As reported earlier, transcriptional activation of cell cycle regulatory target genes is an important component of p53 function. In the present study we analyzed mRNA expression of p53 and p21 which are involved in the control of cell cycle. Our study with colon cancer cells SW480 revealed the activation of p21 although the p53 mRNA level remained unchanged. Immunostaining of ellagic treated SW 480 cells for 48 hrs showed a down regulation of c-myc and proliferative cell nuclear antigen (PCNA). Higher expression of p21 and down regulation c-myc and PCNA in response to ellagic acid exposure further suggests a possible p53 independent and p21 induced cell cycle arrest and apoptosis in SW 480 colon cancer cells. In addition, mRNA expression of the growth factors IGFI and IGFII in ellagic acid treated colon cancer cells showed a down regulation of IGFII, whereas the level of IGFI mRNA was unaffected. The potential action of ellagic acid in preventing cell proliferation through more than one pathway may suggest diverse biological properties of ellagic acid as an anticancer agent at different stages of colorectal tumorigenesis.

Poster Abstract # 17

Dietary Arachidonic Acid Eliminates the Anti-Tumor Effect of Eicosapentaenoic Acid in *Min/+* Mice.

M. Hansen Petrik[a], M. F. McEntee[b], C.-H. Chiu[a] and J. Whelan[a]. Departments of Nutrition[a] and Pathology[b], University of Tennessee, Knoxville, TN, USA

Previously, we demonstrated that arachidonic acid (AA, 20:4 n-6) is the most potent antagonist to eicosapentaenoic acid (EPA, 20:5 n-3) when both are included in the diet (*Li et al., J. Lipid Res 1994*). Several lines of evidence link dietary n-3 fatty acids to a decreased incidence of cancer. This study was designed to investigate the effect of dietary EPA and AA on intestinal tumor load in *Min/+* mice, an animal model which carries a mutation in the human homolog of the tumor suppressor gene *APC*. Twenty male *Min/+* mice were assigned to four dietary groups supplemented with ethyl esters of the following fatty acids (1.5%, w/w), oleic acid (OA, 18:1 n-9), AA, EPA and AA+EPA. After eight weeks on the diets, animals were sacrificed and intestinal fatty acid content, tumor number and size were determined. Supplementing AA increased intestinal 20:4 n-6 content 70%, from 16 mol% to 28 mol% in the OA and AA groups, respectively. EPA supplementation reduced 20:4 n-6 content to 6.5 mol%; however, when AA was added to the diet containing equivalent amounts of EPA (AA+EPA), 20:4 n-6 content increased 3.7 fold to 25 mol% with an 85% reduction in the levels of 20:5 n-3. There were no significant differences in the number and size of tumors among the OA and AA groups. However, with EPA supplementation, tumor number was reduced 68% and 54% compared to the OA and AA groups, respectively, along with a significant reduction in tumor size. When AA was added to the EPA diet (AA+EPA), the anti-tumor effect of EPA was completely eliminated. These data demonstrate that dietary EPA can reduce intestinal tumor load in the *Min/+* mouse model, in part, by reducing tissue AA content. (*This research was funded by grants from AICR, TN-AES, private donations from Myron Pfeifer and supporters of the Arachidonic Acid Project.*)

Poster Abstract # 18

Induction of the transcription factor AP-1 in cultured human colonic cells following exposure to bile acids and human fecal water

Bjorn Glinghammar, Kristina Holmberg, Joseph Rafter

Dept. of Medical Nutrition, Karolinska Institute, Novum, S-141 86 Huddinge, Sweden.

Bile acids are endogenous promotors of colon cancer. The precise mechanisms of their action is to a large extent unknown. The effects of bile acids on cultured colonocytes has been shown to be concentration dependent. At high concentrations they are cytotoxic and at low concentrations they induce proliferation. In HT-29 and HCT 116 cells, physiological concentrations of bile acids induce the protooncogenes c-fos and c-jun, which then complex and form the Activator Protein (AP-1) complex, a transcription factor important in the regulation of cellular growth and malignant transformation. All bile acids induced AP-1 dependent gene transcription, although the effect of the primary bile acid, cholic acid was very weak. Deoxycholic acid (DCA) was the most potent inducer of AP-1 dependent gene transcription. DCA also showed a dose-response induction of AP-1 and this effect was significantly (r=0.91, p=0.01) correlated to the proliferative effect. Human fecal water fractions when added to colonic cells in vitro, cause cytotoxicity. However this effect may not occur in vivo, due to dilution in the protective mucous layer. In order to investigate, at a more physiological concentration, the effects of fecal water on cultured cells, fecal waters were diluted. All fecal waters tested showed an ability to induce proliferation at certain dilutions. At these dilutions, AP-1 dependent gene transcription in the colonocytes was also induced. The results suggest that components of human fecal water, e.g. bile acids, may be contributing to tumor promotion by deregulating the AP-1 transcription factor complex.

Poster Abstract # 19

An Abnormal Receptor for a Vitamin A Derivative in Promyelocytic Leukemia

Robert L. Redner, Elizabeth A. Rush, and Sheri L. Pollock. University of Pittsburgh Medical Center and the University of Pittsburgh Cancer Institute, Pittsburgh PA 15213

Acute Promyelocytic Leukemia is characterized by expression of an abnormal nuclear receptor protein for retinoic acid, a derivative of Vitamin A. The mechanism by which the abnormal receptor produces the leukemic phenotype, and the way in which retinoic acid acts to induce differentiation of the leukemic cells and induce clinical remissions, remains controversial. In an effort to understand this mechanism, we have been studying the t(5;17) variant of promyelocytic leukemia. This translocation fuses the genes encoding nucleophosmin (NPM) and the retinoic acid receptor alpha (RAR). We have developed an in vitro model system for t(5;17) APL by overexpressing NPM-RAR in the monocytic cell line U937. This cell line is capable of differentiating into mature monocytes upon stimulation with Vitamin D3 and TGFß. Subclones expressing NPM-RAR, however, differentiate poorly in response to Vitamin D3 and TGFß. 1 µM all-trans retinoic acid overcomes this differentiation blockade. We have utilized this system to begin to investigate the mechanism of NPM-RAR action. We have produced mutants of NPM-RAR that have a decreased affinity for the RAR heterodimerization partner RXR. When introduced into the U937 cell model, these mutants are less successful than the non-mutated NPM-RAR at blocking Vitamin D3/TGFß mediated differentiation. These results suggest that interaction with RXR is necessary for NPM-RAR induced blockade of differentiation.

Poster Abstract # 20

13-Hydroxyoctadecadienoic Acid (13-HODE) is the Mitogenic Signal for Linoleic Acid-Dependent Growth in Rat Hepatoma 7288CTC *In Vivo**

Leonard A. Sauer, Robert T. Dauchy, David E. Blask, Brenda J. Armstrong and Simone Scalici. Research Institute, The Mary Imogene Bassett Hospital, Cooperstown, NY 13326

Growth of hepatoma 7288CTC in male Buffalo rats is directly dependent on uptake of linoleic acid (LA) from the arterial blood. One to 3% of tumor LA uptake is converted to 13-HODE, an agent that augments EGF-dependent mitogenesis in some cell lines. In this study we tested the hypothesis that LA-dependent tumor growth results from formation of 13-HODE. Tissue-isolated hepatomas (7288CTC) were grown in rats fed either LA-sufficient or –deficient diets. Tumor LA uptake and metabolite release were measured by arteriovenous differences *in vivo* or during perfusion *in situ*. Tumors growing in rats fed a LA-sufficient diet removed LA from the arterial blood and released 13-HODE into the venous blood. Treatment of these rats *in vivo* with the lipoxygenase inhibitor, nordihydroguaiaretic acid (NDGA), caused tumor regression and inhibited tumor 13-HODE release but did not affect LA uptake. Tumors grown in LA-sufficient rats that were perfused *in situ* with donor blood containing LA (1.4 mmole/L plasma) released 13-HODE into the tumor venous blood (0.10 nmol/min/g wet wt) and incorporated [^{3}H]thymidine (42 ± 5 dpm/μg DNA) into tumor DNA. Addition of 13-HODE to the blood perfusate increased [^{3}H]thymidine incorporation to 448 ± 46 dpm/μg DNA; addition of 13-ketooctadecadienoic acid had no effect. NDGA added to the blood perfusate decreased [^{3}H]thymidine incorporation to 21 ± 3 dpm/μg DNA but addition of 13-HODE in the presence of NDGA increased incorporation to 570 ± 200 dpm/μg DNA. Tumors grown in LA-deficient rats that were perfused *in situ* with whole blood from donor LA-deficient rats (plasma LA, 0.3 mM) did not release 13-HODE and [^{3}H]thymidine incorporation was 20 ± 4 dpm/μg DNA. Addition of 13-HODE to the LA-deficient blood perfusate promoted tumor 13-HODE uptake and a dose-dependent increase in [^{3}H]thymidine incorporation. We conclude from these results that 13-HODE is the mitogenic signal responsible for LA-dependent growth in hepatoma 7288CTC *in vivo*.

Poster Abstract # 21

Modulation of Gap Junction-Connexin Gene Expression by the Chemopreventive Bowman-Birk Protease Inhibitor. M.J. Sawey and C.W. Lo. Dept. of Radiation Oncology, Temple Univ. School of Medicine and Dept. Biology, Univ. of Pennsylvania, Philadelphia, PA.

Our preliminary studies suggest that one family of genes that may be targets of regulation by protease inhibitors are those that encode gap junctions membrane channel proteins that mediate cell-cell communication critical in cell growth regulation. The loss or reduction of gap junctional communication has been associated with cell transformation and tumorigenesis. We observed in rodent fibroblast and human mammary epithelial cells that the Bowman-Birk protease inhibitor (BBI) induces the expression of Connexin 43 (Cx43) transcripts, that is associated with an increase in gap junctional communication. Specifically, we observed that treatment of a rodent fibroblast cell line (C3H/10T1/2) with the protease inhibitor BBI, resulted in a 4-5 fold increase in the expression of transcripts from the gap junction gene, Cx43 and that the BBI mediated increase in Cx43 transcripts was correlated with an increase in the level of gap junctional coupling in BBI treated C3H/10T1/2 cells.

We also examined the effects of BBI on Cx43 expression in human mammary epithelial cell lines, MCF-10, MCF-7, and BT-20. MCF-10 is a spontaneously immortalized nontumorigenic mammary epithelial cell line derived from human fibroblastic mammary tissue. In contrast, MCF-7 and BT-20 are tumorigenic cells derived from human breast cancer tissues. BBI treatment of MCF-10 led to an induction of Cx43 transcript expression, while no Cx43 transcripts were detected in MCF-7 or BT-20, either with or without BBI treatment. These findings are particularly intriguing, since in previous studies, BBI has been shown to exert its anticarcinogenic effects only at the initiation and/or promotional stages of tumor progression with no suppressive effects elicited on fully malignant cells or tumors. These observations suggest that gap junction genes may indeed be a target for regulation by the chemopreventive protease inhibitor, BBI.

Poster Abstract # 22

MTHFR polymorphism and risk of colorectal adenomatous polyps

Cornelia Ulrich, Jeannette Bigler, Chu Chen, Ellen Kampman, Stephen Schwartz, Lisa Fosdick, Roberd Bostick, John D. Potter. Fred Hutchinson Cancer Research Center, Seattle, and University of Minnesota, Minneapolis.

The enzyme 5,10-methylene-tetrahydrofolate reductase (MTHFR) may play a role in the etiology of colorectal adenomas via its link to both DNA methylation and nucleotide synthesis. We investigated an association between the common polymorphism (C→T 677) resulting in a thermolabile form of MTHFR, nutrient intake, and colorectal adenomas within the Minnesota case-control study of adenomatous polyps.

Methods Cases (n=527) were diagnosed with colonoscopically confirmed adenomas; controls (n=645) were derived from the same gastroenterology practice, and showed no abnormality at colonoscopy. Dietary intakes were obtained from a self-administered food-frequency questionnaire prior to diagnosis.

Results Age- and sex-adjusted odds ratios (ORs) and 95% confidence intervals for MTHFR status were 0.9 (0.7-1.2) (CT *vs* CC wildtype) and 0.8 (0.6-1.3) (TT *vs* CC). The association between polyps and dietary intakes of folate, vitamin B_{12}, vitamin B_6, or methionine, appeared to differ depending on MTHFR genotype. Individuals with the TT genotype and low intakes of any of these nutrients were at elevated risk for adenomas. An increased risk with increasing alcohol consumption was observed only among those with the CC genotype (p-trend=0.005), whereas among those with the TT genotype, those with moderate alcohol consumption were at lowest risk (p for interaction p=0.02).

Conclusions Nutrient intakes relevant to the MTHFR metabolic pathway appear to modify the relationship between the MTHFR (677 C→T) polymorphism and colorectal adenomas. These data are consistent with a role of DNA methylation and nucleotide synthesis in the pathogenesis of colon cancer. These findings may help to develop more tightly defined prevention strategies.

Poster Abstract # 23

P53 OVEREXPRESSION vs P53 MUTATIONS IN RELATION TO DIETARY RISK FACTORS FOR COLON CANCER.

DW Voskuil[1], E Kampman[1], E Balder[1], GNP van Muijen[2], A van Kraats[2], RA Bausch-Goldbohm[3], P van 't Veer[1]
[1]Division of Human Nutrition and Epidemiology, Wageningen Agricultural University, Wageningen, The Netherlands. [2]Department of Pathology, Nijmegen University Hospital, Nijmegen, The Netherlands. [3]Department of Consumer Research and Epidemiology, TNO Nutrition & Food Research Institute, Zeist, The Netherlands.

Epidemiological studies suggest dietary factors may differently affect p53-dependent and p53-independent pathways to colon cancer. Results of such studies may depend on the method used to detect gene abnormalities. This case-control study of 185 colon cancer cases and 259 population-based controls examines the relation between dietary risk factors and p53 status of tumors, detected by both immunohistochemistry and SSCP(exon 5-8) / sequencing.

Using immunohistochemistry 81 of 185 tumors (44%) were categorized p53 positive, with 20% or more cells stained, and 104 tumors were scored negative. P53 mutations were detected in 59 tumors (32%). Odds ratios (OR) and confidence intervals (95% CI) are calculated using quartile medians as continuous variable, expressed per unit of change which equals the distance between the first and the third quartile. A slight increase in risk of colon cancer was observed for intake of saturated fat when comparing all cases with controls (OR per 16.1 g/day, 1.28; 95% CI, 1.01-1.64). This was largely due to an increased risk in cases without p53 overexpression (1.46; 1.08-1.97), and no association in cases with p53 overexpression (1.07; 0.78-1.47). However, findings were less pronounced when cases were classified by mutation analysis (wildtype: 1.33; 1.01-1.75; mutated: 1.16; 0.81-1.65). Similar results were observed for total fat intake. No differences in p53-dependent and p53-independent pathways were observed for other nutrients nor for vegetable and meat food groups. Interestingly, in cases with transversion mutations in the p53 gene, an increased risk was observed for saturated fat (OR, 2.00; 95% CI, 0.97-4.14), in contrast to those with mutations at CpG sites (0.93; 0.55-1.57). In conclusion, an increase in colon cancer risk for the p53-independent pathway due to fat intake, is more pronounced when using immunohistochemistry. However, mutation analysis is needed to study the possible association with a small group of tumors with transversion mutations.

Poster Abstract # 24

Title: Aspirin supresses the mutator phenotype associated with HNPCC by genetic selection

Authors: S. Wallinger[1], W. Dietmaier[1], T. Bocker[2], R. Fishel[2], J. Rüschoff[1*]

[1]University Regensburg, Institute of Pathology, Regensburg, Germany, [2]Kimmel Cancer Center, Thomas Jefferson University, Philadelphia, USA

Introduction:
Nonsteroidal anti-inflammatory drugs (NSAIDs) have been shown to display cancer preventive and tumor regressive effects in colon cancer. This has been largely attributed to their antiproliferative and apoptosis-inducing activities(1). Most interestingly NSAIDs, particularly sulindac induces regression of adenomas in FAP patients(2). Thus it is very important to understand the mechanisms mediating this phenomenon.

Materials and methods:
Six colorectal cancer cell lines, deficient for a subset of the human mismatch repair (MMR) genes (hMLH1, hMSH2, hMSH6 and hPMS2) were examined on microsatellite instability (MSI) by subcloning. A more rapid detection methodology for MSI in cell lines was developed, that is based on a modified primer extension preamplification (PEP)-PCR technology(3). The cells were treated with different concentrations of aspirin and sulindac over a three months period. MSI frequency was measured at different time intervals in crude cell samples as well as after sorting in apoptotic and non-apoptotic cells.

Results:
The MSI frequency in the MMR-deficient (hMSH2, hMLH1 and hMSH6) cells was markedly reduced after long-term (>10 weeks) exposure to Aspirin or Sulindac. This effect was reversible, time- and concentration-dependent, and appeared independent of proliferation rate and cyclooxygenase function. In contrast, the MSI phenotype of a hPMS2-deficient endometrial cell line was unaffected by Aspirin/Sulindac. We could show that the MSI reduction in the susceptible MMR-deficient cells was confined to non-apoptotic cells, while apoptotic cells remained unstable and were eliminated from the growing population.

Conclusions:
Aspirin/ Sulindac induces a genetic selection for microsatellite stability in a subset of MMR-deficient cells, and may provide an effective prophylactic therapy for Hereditary Non-Polyposis Colorectal Cancer (HNPCC) kindreds where alteration of hMSH2 and hMLH1 genes are associated with the majority of cancer susceptibility cases.

References: 1. Shiff, S.J., et al. J. Clin. Invest. 96: 491(1995); 2. Giovannucci, E., et al. Ann. Intern. Med. 121: 241(1994); 3. Dietmaier, W., et. al. Am. J. Pathol. submitted

contributed by Rüschoff, J., et. al. PNAS in press

INDEX

Aberrant crypt foci, 16, 17, 23, 55, 56, 58, 61, 122
Acetate (acetic acid), 66, 68, 72, 74
Acetyltransferase (NAT1), 96, 97
Adenoma, 87, 91, 99, 101
Adrenal cortical activity, 79
Anchorage, 109
Anoikis, 109, 110
APC and Apc, 10, 15, 16, 18, 20, 23, 31, 46, 100, 102, 106, 108, 109, 110
Apoptosis, 15, 18, 20, 51, 52, 55, 57, 61, 77, 80, 81, 105, 106, 107, 109, 110, 112, 123
Arachidonic acid, 119, 124
Armadillo, 24, 25
Aspirin, 34, 58, 60, 91, 120
ATP (adenosine triphosphate), 18
Attenuated familial polyposis (AAPC), 102
Axin, 28, 30
Azoxymethane, 36, 37, 56, 57, 59, 61, 106, 109, 111, 112

Bad, 108
Bak, 108
Bax, 20, 106, 108, 109
Bcl, 20, 106, 108, 109, 110
Bir, 108
BRAG-1, 108
Bran, 65, 68, 70, 71, 74, 90
Butyrate (butyric acid), 17, 20, 66, 68, 70, 74, 108, 110

Cadherin, 10, 109, 110
Calcium, 73, 89, 90, 92
Cancer, colon, 33, 55, 56, 85, 87, 100
2-Carboxyphenylretinamide (2-CPR), 56, 57, 60, 61
Carcinogenesis, 77, 79, 106
Carotene, beta, 89, 92
Caspase, 17, 18, 108
Catenin, 18, 20, 23, 31, 100, 106, 109, 110
Cdc, 120, 123
Cdx, 9, 10, 11
Cellulose, 66, 70, 111, 112
Colectomy, 47
Colon, apoptosis, 15, 106, 107
  cancer, 33, 85, 95, 99, 102, 105, 119
  cell proliferation, 15, 17, 105, 107
  differentiation, 15, 19, 111

Concavalin A, 41
Cortical steroids, 78, 79, 81
Crypts, intestinal, 2, 4, 8, 16, 65, 105, 107, 109, 111
Cycle, cell, 65, 66, 77, 100, 102, 119, 121
Cyclins and cyclin-direct kinases (cdks), 17, 65, 67, 69, 70, 73, 81, 82, 120,121, 123, 124
Cyclooxygenases (COX), 47, 54, 106, 110, 119, 121, 124
Cytokines, 34, 38

Development, colonic epithelium, 1, 2, 4
Diet, 34, 35, 85, 87, 105
Dietary factors, 86, 110
  intervention, 85, 89, 91, 92, 95
  restriction, 77, 79, 81, 82, 110
Differentiation, 15, 16, 19, 107
a-Difluoromethylornithine (DFMO), 59, 60
Dimethylhydrazine, 46
DNA, adducts, 112
  damage, 109
  repair, 108
Downregulation, 25, 28, 30
Drosophila, 24, 25

ELISA, 39
Endoderm, 6, 7
Energy restriction, 77, 82
Enterocytes, 2
Enteroendocrine cells, 2
Epidermal growth factor (EGF), 25
Estrogen, 80
Exisulind, 45, 47, 48, 51, 53

Familial adenomatous polyposis (FAP), 45, 46, 51, 53, 88, 99, 102, 103
Familial clustering, 101, 103
Fas-L, 108
Fat, dietary, 85, 86, 91, 92, 110, 111
Fatty acids, 33, 37, 41, 110, 111
Fecal bile acids, 73
FGN-1 (sulindac sulfone), 48, 49, 51, 52
Fiber, 65, 66, 90, 92, 110, 111
Fibroblasts, intestinal, 7
Folate, 89

Forskolin, 17
Free radicals, 34

Gastrin, serum, in colon cancer, 103
Genetics, colorectal cancer, 95
 polymorphisms, 96, 102
 susceptibility, 95
Genes, fatty acid binding proteins, 3
 lactase-phlorizin hydrolase, 3
 sucrose-isomaltase, 3
 zeste white, 3, 25
Glutathione-S-transferase, 96, 97
Glycogen synthase kinase (GSK3β), 24, 25, 27, 30, 70
Goblet cells, 6, 19
Growth arrest, 66

HCT cells, 124
Hepatocyte nuclear factor, 9
Hepatotoxicity, 45
Hereditary cancer, 95
 nonpolypotic colon, 95
 nonpolyposis colorectal, 95, 96, 99, 100, 102, 103
Histone deacetylase, 20
Homeobox genes, 1
Homeostasis, 15, 16, 23, 77, 81, 105
Human, colon carcinoma cell line (U4), 65, 71
 HT-29 colon cancer line, 10, 19, 120, 122, 124
 intestinal development, 2, 4
 kidney cell line, 23, 25, 26, 29
4-hydroxyphenylretinamide (4-HPR), 56, 57, 60

IEC-b (rat intestinal cell line), 10, 38, 40
Image processing, 17
Immune system, 33, 40, 41, 79
Indomethacin, 34, 46, 120
Inflammation, 33, 35
Integrin, 74
Intervention, 85, 87
Intestine, polyps, 46, 50
Iron, 33, 35, 40

Ki-67, 46, 48, 50, 52, 123
K-ras, 55

Lipids, dietary, 33, 35, 37, 40, 42
Lymphocytes, colonic, 40, 41

Macronutrients, supplements, 90
Mad, 108
Mammary cancer, 78, 80, 82
Manganese, mucosal, 37
 superoxide dismutase (MnSOD), 34, 40, 42
Mcl, 108
Meat, 86
Mesenchyme, 4
Mesoderm, 6, 7
Micronutrients, supplements, 88, 90
Microsatellite instability, 95, 100
Mismatch repair, 96, 97, 100, 102

Mitochondrial, genome, 17
 membrane potential, 18, 20
MLH1, 96
Models, rodent for colon tumors, 107
Monoclonal antibody, 47
Mouse, intestinal development, 3, 5
MSH2, 96, 100
MUC2, gene, 19
Mucin, 19, 46, 52
Mutation, 96, 100, 102, 106
 cluster region (MCR), 25
Myc, 15, 18, 20, 25, 29, 108

Naproxam, 120
Nonsteroidal anti-inflammatory agents (NSAIDs), 34, 45, 46, 55, 61, 110, 119, 120, 125

Oil, dietary, corn, 36, 37, 41, 111, 112
 fish, 36, 37, 41, 111, 112
Oncogene activation, 23, 55
Oncogenesis, 33
Ornithine decarboxylase, 59

P21, 18, 66, 73, 121, 124
P27, 66, 67, 69, 73, 77, 81, 82
P53, 15, 52, 65, 107, 109, 121, 122
P57, 67
PRb, 121, 123
Paneth cells, 2
Pectin, 66, 70, 71, 73, 111, 112
Phentlethyl-3-methylcaffeate, 61
Phosphorylation, 27, 31, 66
Piroxicam, 57, 58, 60, 120, 124
Polyps, colonic, 46, 50, 53, 91, 99, 100, 102
Polyunsaturated fatty acids (PUFA), 110, 113
Prevention trials, 87
Proliferating cell nuclear antigen (PCNA), 56, 87, 123
Proliferation, 16, 48, 51, 52, 55, 59, 61, 77, 79, 80, 81, 105, 106
Propionate (propionic acid), 66, 68, 72, 74
Prostaglandin, 41, 42, 51, 119
Protein kinase, 19, 111, 113

Quercetin, 61

Ras, 15, 65, 123
Regression, rectal polyps, 45
Regulation, cell cycle, 120
Replication, errors, 100
Retinoic acid, 9-cis, 56, 57, 60
Retinoids, 55, 56, 61
Risk, familial, 101, 103
RNA, polymerase and synthesis, 9

Salicylic acid, 124
Scanning, 17
Selenium, 89
Short chain fatty acids (SCFAs), 16, 19, 65, 73
Signaling, 25, 123

Sulindac, 34, 45, 46, 51, 120, 122
Sulindac sulfide, 47, 122, 124
Sulindac sulfone (Exisulind), 45, 47, 48, 51, 53, 120
SW480, colorectal cancer cells, 25, 26

T-cells, 40
Thromboxane, 119
Tissue, size, 79, 80
TPA (tetradecanoylphorbol acetate), 19
Transcription and factors, 8, 9, 10, 24, 27, 66, 100
Transgenic mice, 8, 9
Transforming growth factor (TGF), 65, 68, 71, 74, 96, 108, 120
Trichostatin A, 20
Tumor necrosis factor (TNF), 37, 38, 40
Tumors, initiation and promotion, 56, 57, 59, 106
TUNEL assay, 16, 46, 48, 123

U4 cell line (human colon carcinoma), 65, 72

Valinomycin, 18
Villi, intestinal, 2, 4, 6, 8
Vitamins (A, C, D, E), 89

WAF1/Cipl, 107
Wnt-1, 24, 28, 29

Xenopus, 24, 25, 28

FSC
www.fsc.org
MIX
Papier aus verantwortungsvollen Quellen
Paper from responsible sources
FSC® C105338

If you have any concerns about our products,
you can contact us on
ProductSafety@springernature.com

In case Publisher is established outside the EU,
the EU authorized representative is:
Springer Nature Customer Service Center GmbH
Europaplatz 3, 69115 Heidelberg, Germany

Printed by Libri Plureos GmbH
in Hamburg, Germany